XLIV Zvenigorod International Conference on Plasma Physics and Controlled Fusion (ICPAF 2017)

Journal of Physics: Conference Series Volume 907

Zvenigorod, Russia
13 - 17 February 2017

Editors:

V.A. Ivanov
I.A. Grishina
E.V. Voronova

ISBN: 978-1-5108-5038-5
ISSN: 1742-6588

Printed from e-media with permission by:

Curran Associates, Inc.
57 Morehouse Lane
Red Hook, NY 12571

Some format issues inherent in the e-media version may also appear in this print version.

Printed by Curran Associates, Inc. (2017)

For permission requests, please contact the Institute of Physics at the address below.

Institute of Physics
Dirac House, Temple Back
Bristol BS1 6BE UK

Phone: 44 1 17 929 7481
Fax: 44 1 17 920 0979

techtracking@iop.org

Additional copies of this publication are available from:

Curran Associates, Inc.
57 Morehouse Lane
Red Hook, NY 12571 USA
Phone: 845-758-0400
Fax: 845-758-2633
Email: curran@proceedings.com
Web: www.proceedings.com

Table of contents

Volume 907

XLIV Zvenigorod International Conference on Plasma Physics and Controlled Fusion (ICPAF2017)

13–17 February 2017, Zvenigorod, Moscow region, Russian Federation

Accepted papers received: 6 October 2017
Published online: 22 October 2017

Preface

Papers

IOP Conf. Series: Journal of Physics: Conf. Series **907** (2017) 011001 doi:10.1088/1742-6596/907/1/011001

Introduction

This issue of *Journal of Physics: Conference Series* includes 28 articles submitted by the participants of the XLIV International Zvenigorod Conference on Plasma Physics and Controlled Fusion (February 13-17, 2017). The conference is named after the town of Zvenigorod, Moscow region, where it has been annually held for 44 years. More than 600 scientists took part in the conference this year. These articles showcase a wide range of topical issues in the field of plasma physics that were discussed at the conference.

Leading foreign scientists took part even in the first Zvenigorod conferences. They were invited by the conference organizers to give talks on the challenging problems of plasma physics. By now, the conference has truly become international. Several dozens of foreign scientists from the USA, Europe, and Japan take part in the conference annually. They represent about 30 leading research centers working in different fields of plasma physics.

Nowadays, the conference has expanded to include a large number of topics. The conference schedule includes sessions on magnetic confinement of high-temperature plasma, inertial fusion, processes in low-temperature plasma, physical principles of plasma and beam technologies, and topical issues of the ITER project. About 90 Russian research centers are usually represented at the conference. Among them are institutes of the Russian Academy of Sciences, research centers of Russian universities, institutes of the State Corporation ROSATOM, and other scientific organizations of the Russian Federation.

The conference talks are selected by the Programming Committee comprised of leading Russian and foreign scientists.

Leading scientists from around the world working in the field of plasma physics have become members of the International Advisory Board of the International Zvenigorod Conference on Plasma Physics and Controlled Fusion. The International Advisory Board is aimed to ensure and confirm the international status of the conference, which takes place in Russia.

The full list of papers can be viewed on the conference website http://www.fpl.gpi.ru/Zvenigorod/XLIV/Zven_XLIV_e.html.

The issue was edited by V.A. Ivanov, I.A. Grishina, and E.V. Voronova.

Peer review statement

All papers published in this volume of *Journal of Physics: Conference Series* have been peer reviewed through processes administered by the proceedings Editors. Reviews were conducted by expert referees to the professional and scientific standards expected of a proceedings journal published by IOP Publishing.

Experimental studies of toroidal correlations of plasma density fluctuations along the magnetic field lines in the T-10 tokamak and first results of numerical modeling

M A Buldakov[1], V A Vershkov, M Yu Isaev and D A Shelukhin

NRC "Kurchatov Institute", 123182, pl. Kurchatova 1, Moscow, Russia

[1] E-mail: m.a.buldakov@gmail.com

Abstract. The antenna system of reflectometry diagnostics at the T-10 tokamak allows to study long-range toroidal correlations of plasma density fluctuations along the magnetic field lines. The antenna systems are installed in two poloidal cross-sections of the vacuum chamber separated by a 90° angle in the toroidal direction. The experiments, which were conducted at the low field side, showed that the high level of toroidal correlations is observed only for quasi-coherent fluctuations. However, broadband and stochastic low frequency fluctuations are not correlated. Numerical modeling of the plasma turbulence structure in the T-10 tokamak was conducted to interpret the experimental results and take into account non-locality of reflectometry measurements. In the model used, it was assumed that the magnitudes of density fluctuations are constant along the magnetic field lines. The 2D full-wave Tamic-RTH code was used to model the reflectometry signals. High level of correlations for quasi-coherent fluctuations was obtained during the modeling, which agrees with the experimental observations. However, the performed modeling also predicts high level of correlations for broadband fluctuations, which contradicts the experimental data. The modeling showed that the effective reflection radius, from which the information on quasi-coherent plasma turbulence is obtained, is shifted outwards from the reflection radius by approximately 7 mm.

1. Measurements of toroidal correlation of density fluctuations along magnetic field lines in the T-10 tokamak

Correlation reflectometry experiments, which have been previously conducted in the T-10 tokamak, have shown [1, 2] that the plasma turbulence spectrum can be represented as a sum of several types of density fluctuations, which have different radial, poloidal and toroidal correlational properties. The antenna arrays of T-10 tokamak enabled to conduct measurements of correlations of plasma density fluctuations along the magnetic field lines. The results of the first experiments on toroidal correlations of plasma density fluctuations along the field lines have been published in [1–4] for the different antenna arrays. This article presents the first results on correlations of plasma density fluctuations along the field lines with a new antenna configuration. Such experiments can provide additional information about the physical properties of different types of plasma fluctuations in tokamak, e.g., their correlational lengths along the magnetic field lines. Furthermore, these experiments can provide additional information about the q profile and allow to estimate the electric current profile in a tokamak plasma column. Schematic of the experiment for measuring correlations of plasma density fluctuations along the magnetic field line is shown in figure 1. There are antenna arrays in the cross-section D at LFS and in the cross-section A, which is separated in the toroidal chamber of tokamak by

IOP Conf. Series: Journal of Physics: Conf. Series **907** (2017) 012001 doi:10.1088/1742-6596/907/1/012001

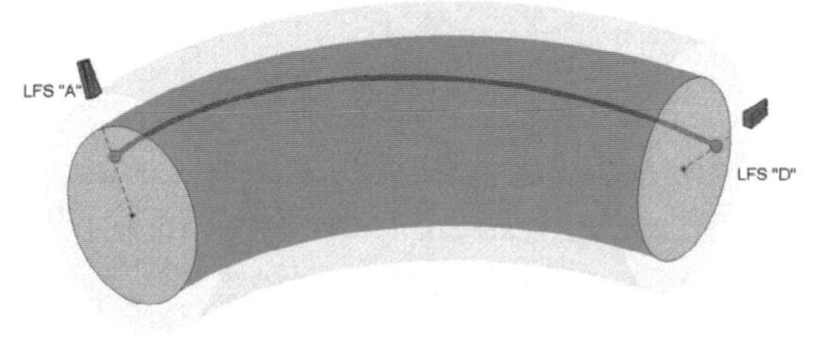

Figure 1. Schematic of the experiment on measuring correlations of plasma density fluctuations along the magnetic field line. The field line is represented by the solid line.

90° from the cross-section D at LFS. The antenna arrays are installed in cross-sections D and A at poloidal angles 0° and 68.5°, respectively. The antenna arrays installed at toroidally separated cross-sections allow to conduct measurements of toroidal correlations of plasma density fluctuations at LFS. The experiments were carried out for the combination of antennas, which are shown in figure 1.

The experiments were conducted using an O-mode heterodyne reflectometer at a constant frequency. The parameters of discharge were as follows: toroidal magnetic field B = 2.4 T, toroidal plasma current I = 250 kA. In the experiments, plasma was probed via reflectometry at two ends of the magnetic field line, and correlations between the two antenna signals were calculated. The reflection radius and corresponding q values were changed by varying the reflectometer frequency. The characteristic spectra and correlational characteristics of signals [5] of two antennas are shown in figure 2.

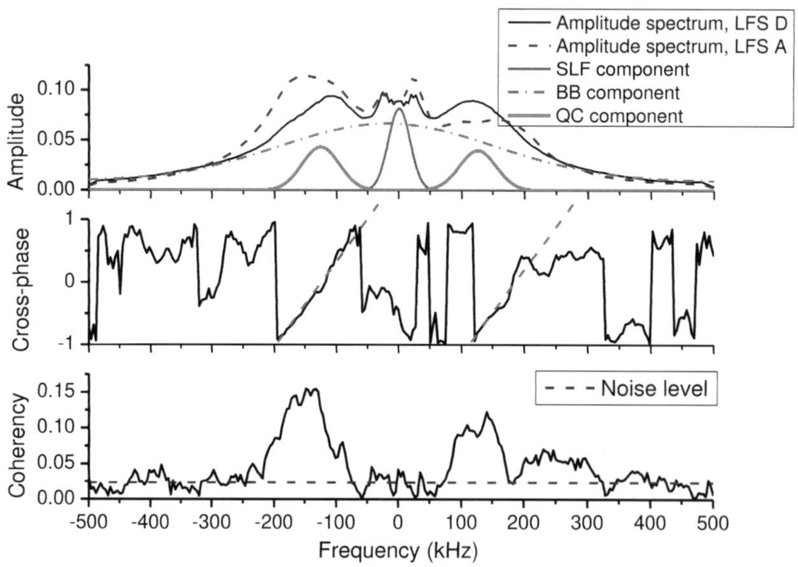

Figure 2. Amplitude spectra (top), cross-phase (middle) and coherency (bottom) of electric field fluctuations at the receiving antennas in the experiments. Decomposition of the spectrum into stochastic low frequency (SLF), broadband (BB) and quasi-coherent (QC) fluctuations is also shown.

Coherency between the signals, which is several times higher than the noise level, is observed only at the [100–200] kHz frequency band, which corresponds to quasi-coherent (QC) fluctuations. Thus, the experiments showed that toroidal correlations of plasma density fluctuations are observed only for the QC fluctuations. Stochastic low frequency (SLF) and broadband (BB) fluctuations are not correlated at the distance of 2.5 m along the magnetic field line (¼ turn of the torus circumference). The slope of the phase, which is represented by the dashed line in the middle figure, corresponds to the delay of 7 μs between the two signals. In the case shown in figure 2, reflection occurs not at the resonant radius, where delay time should be equal to zero.

2. Numerical modeling of the plasma turbulence structure and reflectometry signals in the T-10 tokamak

Numerical modeling of the turbulence structure in the T-10 tokamak was conducted to interpret the experimental data and take into account non-locality of reflectometry. The 2D stochastic model of turbulence was used [5]. Broadband and one type of quasi-coherent fluctuations were modeled. Fluctuations were generated on the calculation grid at the LFS cross-section D, and it was assumed that the amplitudes of plasma density fluctuations are constant along the magnetic field lines. The structure of fluctuations was mapped along the field lines to the area of antennas at the LFS cross-section A. Antenna signals at the cross-sections A and D at LFS were modeled using the 2D full-wave electromagnetic code Tamic-RTH for the ordinary wave [5]. The modeling results are shown in figure 3.

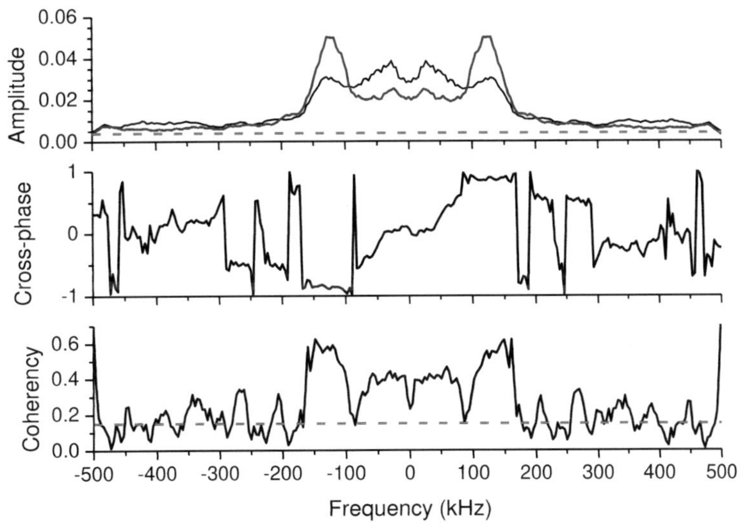

Figure 3. Amplitude spectra (top), cross-phase (middle) and coherency (bottom) of the electric field fluctuations at receiving antennas in the model. The noise level is shown by the dashed lines.

High level of coherency above the noise is observed at the [0–100] kHz (BB fluctuations) and [100–200] kHz frequency bands (QC fluctuations). Thus, for the modeled signals, toroidal correlations are high for the QC fluctuations, which agrees with the experiments. High toroidal correlations are also observed for the BB fluctuations ([0–100] kHz), which contradicts the experimental data.

Additional information about the plasma density fluctuations can be obtained by analyzing the delay time Δt between fluctuations of the signals of the two antennas. Because the geometry of antennas was fixed, the delay time Δt is supposed to be equal to zero at the resonant radius, i.e., the radius of magnetic field line that connects reflection points of the signals of the two antennas. If the reflection radius differs from the resonant radius, non-zero delay should be observed due to poloidal rotation of plasma turbulence. The dependence of delay time on reflection radius for the modeled reflectometry signals and for the local values of plasma density fluctuations on the same radii were compared. Those dependences are shown in figure 4. Delay time was calculated for the quasi-coherent turbulence.

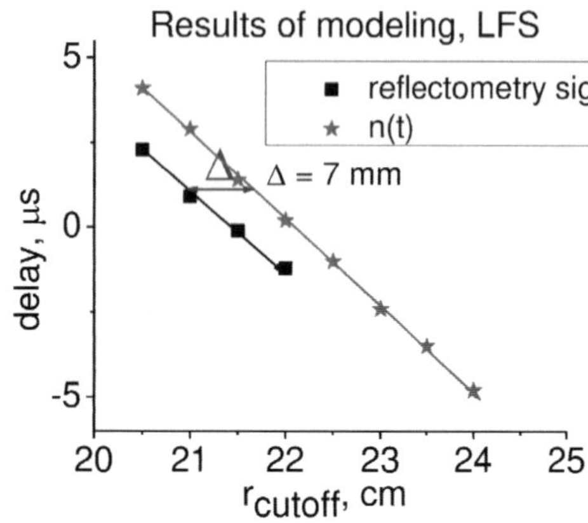

Figure 4. Dependence of delay time on cut-off radius for the modeled reflectometry signals (squares) and corresponding local values of density fluctuations (asterisks).

It can be seen from figure 4 that there is a radial shift $\Delta \approx 7$ mm between the graphs for the modeled reflectometry signals and for the local values of density fluctuations. This can be explained by the fact that the reflectometry signal acquires phase along its path of propagation, and its total phase incursion depends on the distribution of plasma density fluctuations and its sensitivity along the radius. This leads to the shift between the reflection radius and the effective radius where the most information about plasma density fluctuations is acquired. This shift Δ should be taken into account in the experiments on toroidal correlation measurements of plasma density fluctuations along the field lines.

3. Conclusions

The paper presents the first experimental and modeling results on the measurements of long distance correlations of plasma density fluctuations along the field line with a new antenna system at LFS. High correlations along the magnetic field lines are experimentally observed only for the QC fluctuations. The SLF and BB fluctuations are not correlated at the distance of 2.5 m. The high level of correlation of QC fluctuations in the experiments and modeling indicates that the assumption that the amplitudes of plasma density fluctuations are constant along the field lines is true for this type of turbulence. However, the modeling showed that toroidal correlations of plasma density fluctuations are also observed for the BB fluctuations ([0–100] kHz), which does not agree with the experiments. This shows that the assumption that the amplitudes of fluctuations are constant along the field lines is incorrect for this type of turbulence, at least for the experimentally investigated distance 2.5 m. The question of BB fluctuations properties has not been solved yet, and further research is needed. The modeling also showed that the effective radius, which contains information about the QC turbulence, is shifted from the reflection radius outwards by $\Delta \approx 7$ mm due to non-locality of reflectometry. This shift should be taken into account in the local measurements of the current profile using rotational transform measurements.

Acknowledgments

This work was supported by Rosatom State Corporation (contract H.4x.241.9Б.17.1011). Correlational measurements were supported by the Russian Science Foundation project 14-22-00-193.

References

[1] Vershkov V A et al 1999 *Rev. Sci. Instrum.* **70** 1700-09

IOP Conf. Series: Journal of Physics: Conf. Series **907** (2017) 012001 doi:10.1088/1742-6596/907/1/012001

[2] Vershkov V A et al 2005 *Nuclear Fusion* **45** S203-26
[3] Vershkov V A, Soldatov S V, Shelukhin D A and Chistyakov V V 1999 *Nucl. Fusion* **39** 1775
[4] Vershkov V A 2007 *Recent Research Developments in Plasma Physics* ISBN: 978-81-7895-307-6 (Kerala, India: Transworld Research Network) 119-194
[5] Urazbaev A O et al 2006 *Plasma Phys. Reports* **32** 443-60

Measurement of geodesic acoustic modes and the turbulent particle flux in the T-10 tokamak plasmas

L G Eliseev[1], V N Zenin[1], S E Lysenko[1,3] and A V Melnikov[1,2]

[1] NRC "Kurchatov Institute", 123182 Moscow, Russia
[2] National Research Nuclear University MEPhI, 115409 Moscow, Russia

[3] E-mail: Lysenko_SE@nrcki.ru

Abstract. Geodesic acoustic modes (GAMs) and the turbulent particle flux were studied in the T-10 tokamak in discharges with Ohmic (OH) and electron cyclotron heating (ECRH). The broadband oscillations of electric potential and density with frequencies up to 250 kHz were measured using a heavy ion beam probe (HIBP) in the plasma core. At the plasma edge, at a relative radius $\rho > 0.8$, the dominant GAM peak with a frequency of approximately 20 kHz and the noticeable peak of quasicoherent oscillations in the 40–100 kHz range were observed. Using a multislit energy analyzer, we estimate the poloidal electric field E_{pol} and the radial electrostatic turbulent particle flux, which is driven by drift in crossed $E \times B$ fields. The GAM peak was observed in the spectrum of potential oscillations, but it was practically absent in the spectrum of E_{pol} oscillations and in the frequency-resolved particle flux. However, the flux was observed in the broadband range. These results agree with the theoretical concept of GAM and our earlier measurements of symmetric poloidal structure of GAM potential perturbations.

1. Introduction

Zonal flows (ZFs) and their high-frequency counterpart, the geodesic acoustic modes (GAMs), are considered as a possible mechanism of the high-temperature plasma turbulence self-stabilization [1]. Therefore, ZFs and GAMs have been intensively studied in many fusion devices, see review [2]. According to modern theoretical concepts, turbulent transport has a complicated structure caused by the interplay of fine-scale turbulence (e.g., ITG, TEM and other modes) via middle-scale turbulence (ZF and GAM) with global events (e.g., L-H transitions) [3]. The experiments and theory show that GAMs convert the radial turbulent flux, which exhausts the plasma energy to the wall, into the torsional oscillations of potential [4], which do not transfer energy to the wall. GAMs are driven by the broadband turbulence, and after its depletion, GAMs are saturated and die from 'starvation'. So, GAMs have an oscillating character, and their interplay with turbulence is dubbed as 'prey-predator'.

The radial turbulent particle flux Γ_r is a product of density n_e and radial velocity V_r oscillations, which are averaged over some frequency domain:

$$\Gamma_r = \left\langle \tilde{n}_e \cdot \tilde{V}_r \right\rangle . \tag{1}$$

In turn, V_r is driven by the $E \times B$ drift that is linked to the poloidal electric field E_{pol} and toroidal magnetic field B_t as follows:

$$\tilde{V}_r = \tilde{E}_{pol} / B_t . \tag{2}$$

The paper presents first measurements of the GAM dynamics together with the radial turbulent particle flux at the T-10 tokamak. These measurements are possible only in the core plasma using a unique diagnostic tool, a heavy ion beam probe (HIBP).

2. Experimental setup

Experiments were performed in the T-10 tokamak (major and minor radii are $R = 1.5$ m and $a = 0.3$ m, respectively, toroidal magnetic field is $B_t = 1.6$–2.4 T, plasma current is $I_p = 0.15$–0.3 MA, line-average plasma density is $\bar{n}_e = (0.6–5) \times 10^{19}$ m^{-3}. The shots with Ohmic and electron cyclotron resonance heating (ECRH), with a power $P_{EC} = 0.5$ MW were studied.

The T-10 tokamak is equipped with a multichannel heavy ion beam probe (HIBP) [5], which measures the oscillating components of density n_e and potential φ in a wide radial interval, $+0.2 < \rho < 1$, where $\rho = r/a$. HIBP operates as follows. The primary beam of Tl^+ ions with the energy $E_{b1} = 200$–300 keV is injected into the plasma at an entrance angle α and moves along the Larmor orbit. At some point (sample volume or *SV*), the Tl^+ ions are ionized, and the secondary current I_{tot} of Tl^{++} ions with the energy E_{b2} moves along another orbit and reaches the energy analyzer. The local potential in *SV* is equal to the energy difference: $e\varphi = E_{b2} - E_{b1}$, while the local density is proportional to I_{tot}. To estimate the absolute level of density fluctuations, we need to calibrate $\tilde{I}_{tot}$ using the interferometer data. The T-10 energy analyzer has 5 entrance slits. Now, we use the lateral and central slits, which are denoted as S_1, S_3 and S_5. Each slit observes its *SV*. The *SV* position depends on E_{b1}, α, B_t, and device geometry. The set of *SVs* forms the detector grid. A sketch of the experiment and the points of detector grid are shown in figures 1 and 2, respectively. If we measure the potentials in *SVs* that are observed by slits S_1 and S_3, which are located at the same flux surface at a distance Δx, then, $E_{pol} = (\varphi_1 - \varphi_3)/\Delta x$. In reality, the line S_1–S_5 does not coincide with the flux surface. Thus, apart from E_{pol}, we measure the radial component E_r. Therefore, we try to choose such position of *SVs*, where $E_{pol} >> E_r$.

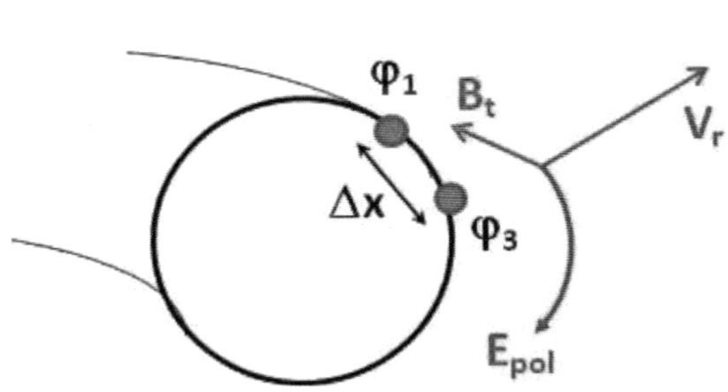

Figure 1. Sketch of the electric field E_{pol} and radial velocity V_r measurements; φ_1 and φ_3 are the potentials of sample volumes on the same flux surface, which are measured by slits S_1 and S_3, respectively, at a distance Δx = 1.2 cm.

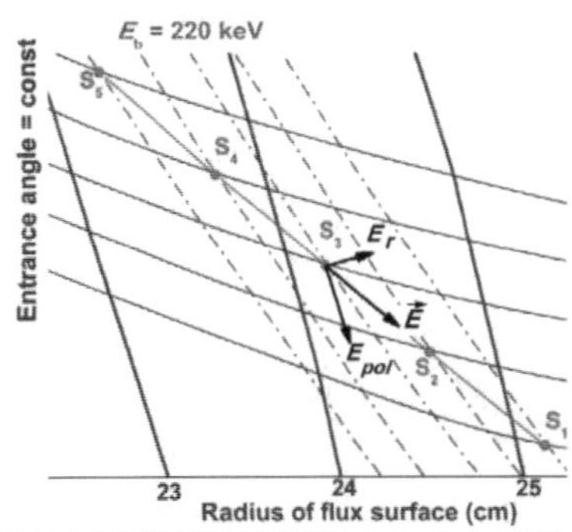

Figure 2. Spitted point of the detector grid. Flux surfaces (thick lines) do not coincide with the line of sample volumes (S_1–S_5). Thus, HIBP measures a mixture of poloidal E_{pol} and radial E_r components of the electric field.

3. Experimental results

After processing the HIBP signals in the frequency range of less than the Nyquist frequency, $f_N \sim 250$ kHz, we find their Fourier-transforms or power spectral densities *PSD* for $\tilde{n}_e$ and $\tilde{V}_r$, their coherence $coh(\tilde{n}_e, \tilde{V}_r)$, cross-phase, $\theta = \text{Arg}(F_n, F_V^*)$, (* means the complex conjugate), cross-spectral density $CSD(\tilde{n}_e,\tilde{V}_r)$, flux spectral function $\mathrm{Re}\,CSD(\tilde{n}_e,\tilde{V}_r)$ and the total frequency-averaged particle flux.

$$\Gamma(t) = (1/f_N)\int_{f_1}^{f_2} \mathrm{Re}(F_{n_e}(f,t)\cdot F_{V_r}(f,t)^*)\mathrm{d}f \qquad (3)$$

Figure 3 presents PSD for the potential φ_3 (*a*), electric field $E_{pol} \sim \varphi_1 - \varphi_3$ (*b*) and the spectral function of flux (*c*) fluctuations measured at $\rho = 0.57$. At this radius, the high-frequency broadband turbulence is absent. Therefore, we show only the low-frequency part ($f < 60$ kHz) of the spectra. The GAM peak, which is marked by a black rectangle, is visible on the spectrum of potential, but it is practically absent on the spectrograms of E_{pol} and the flux. Note that at $t > 650$ ms, the large-scale instabilities are seen on the density (*c*) and all HIBP signals.

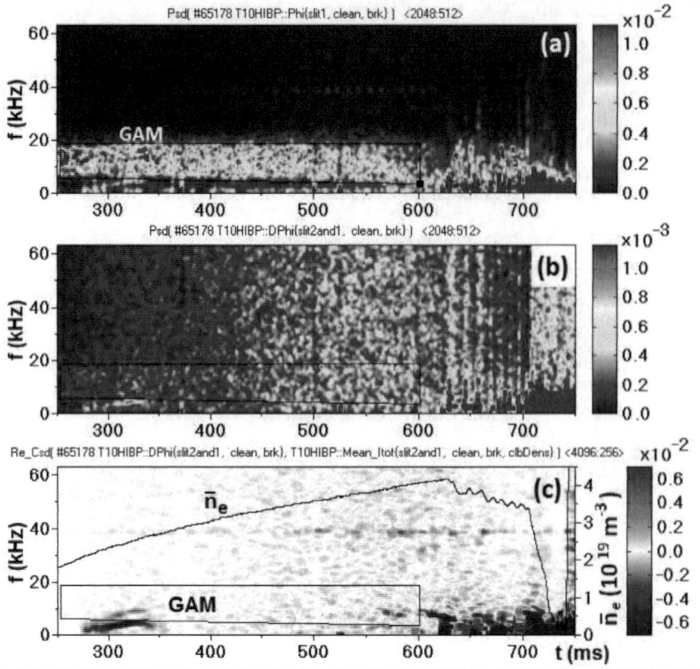

Figure 3. (*a*) *PSD* of the potential φ_3; (*b*) *PSD* of $E_{pol} \sim \varphi_1 - \varphi_3$; (*c*) flux spectral function Re $CSD(n_e, V_r)$ measured at $\rho = 0.57$, and the line-average density $\bar{n}_e$.

For the quasicoherent (QC) oscillations ($100 < f < 170$ kHz), the coherence between E_{pol} and density n_e is rather high, $coh > 0.6$, the phase between E_{pol} and n_e is $\theta \sim 1.5$–3 rad and increases with frequency. Thus, the intense flux, which is directed outward, is observed for QC, figure 4. In this shot, we also see the weak flux in the GAM frequency range [6]. For this shot, we estimated the root mean square of E_{pol}, RMS(φ_1-φ_3) = 13 V, RMS of the flux

$$(\tilde{\varphi}_1 - \tilde{\varphi}_3)\cdot\frac{\tilde{I}_{tot}}{\bar{I}_{tot}} = \frac{\Gamma_r B_t \Delta x}{n_e} \approx 0.05\ \mathrm{B}, \qquad (4)$$

where $n_e(\rho = 0.78) = 2.3 \times 10^{19}$ m^{-3}. Therefore, the total specific flux averaged over the QC frequency range is $\Gamma_r(0.78) \leq 4 \times 10^{19}$ m^{-2} s^{-1}. The HIBP measurement for a similar shot gives $\Gamma_r(\rho = 0.87) \leq 2 \times 10^{20}$ m^{-2} s^{-1}, while the Langmuir probe in a different regime gives $\Gamma_r(\rho = 1) \leq 2 \times 10^{20}$ m^{-2} s^{-1} [7], i.e., the flux increases with radius. If we suppose the poloidal and toroidal homogeneity of the flux, and multiply $\Gamma_r(0.87)$ by the lateral surface of T-10, $S = 2\pi R\cdot 2\pi a$, we obtain the total outflow $Q_r = 3 \times 10^{21}$ s^{-1}, which is comparable with the total particle influx, or the Mukhovatov number $N_M = (2.5–3) \times 10^{21}$ s^{-1} [8].

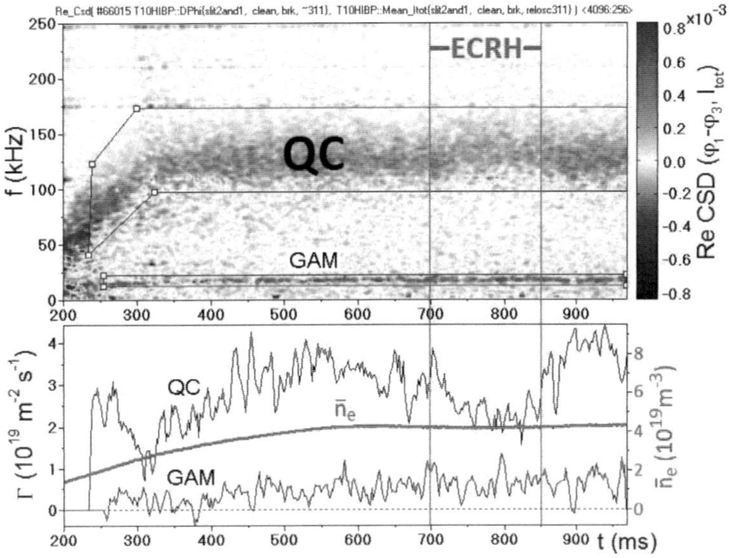

Figure 4. (*a*) Flux spectral function; (*b*) total particle fluxes averaged over the GAM and QC frequency ranges measured at $\rho = 0.78$, and the line-average density $\bar{n}_e$.

4. Conclusions

Using HIBP with a multislit analyzer, we performed the first simultaneous measurements of the potential, electric field and density oscillations at the core and edge of tokamak plasmas, and their correlations. We estimate that the radial turbulent particle flux is driven by the $E \times B$ drift. The absence of flux in the GAM frequency range agrees with a theoretical concept of GAM, as the high-frequency branch of zonal flow with a symmetrical poloidal structure of potential perturbations [1], and with our previous measurements of the poloidal mode number $m = 0$ for GAM [9, 10]. The broadband quasicoherent turbulence of plasma forms a considerable part of the total particle outflow.

Acknowledgments

The work is carried out by the Russian Scientific Foundation, project 14-22-00193. The T-10 experiments were partly supported by the State Corporation Rosatom contract No H.4x.241.9B.17. The work of AVM was partly supported by the Competitiveness Programme of NRNU MEPhI.

References

[1] Diamond P H *et al* 2005 *Plasma Phys. Control. Fusion* **47** R35
[2] Fujisawa A 2009 *Nucl. Fusion* **49** 013001
[3] Miki K and Diamond P H 2011 *Nucl. Fusion* **51** 103003
[4] McKee G R *et al* 2003 *Plasma Phys. Control. Fusion* **45** A477
[5] Melnikov A V *et al* 2017 *Nucl. Fusion* **57** 072004
[6] Melnikov A V *et al* 206 *Plasma Phys. Control. Fusion* **48** S87
[7] Solomatin R Y *et al* 2015 *42nd EPS Conf. Plasma Physics*, Rep. P5.160

[8] Vershkov V A *et al* 2012 *Probl. Atom Sci. Techn.*, ser. Thermonuclear Fusion (4) 8 (in Russian)
[9] Zenin V N *et al* 2014 *Probl. Atom Sci. Techn., ser. Plasma Physics* **20** 269
[10] Melnikov A V *et al* 2015 *Nucl. Fusion* **55** 063001

Co- and counter-pellet-injection in T-10 tokamak

V G Kapralov[1,3], Yu D Pavlov[2], A E Borovov[1], M M Dremin[2], S V Krylov[2], D V Ryzhakov[2], A S Trubnikov[2], V G Skokov[1], V V Elagin[1], H A Harfush[1] and K S Sedov[1]

[1] Peter the Great St. Petersburg Polytechnic University, St. Petersburg, RF
[2] National Research Center «Kurchatov institute», Moscow, RF

[3] Email: kapralov@phtf.stu.neva.ru

Abstract. The article presents the results of experimental studies and modelling of chord pellet injection in the T-10 tokamak plasmas. The deviation of pellets at different angles up to the peripheral injection does not result in their substantial degradation for speeds of approximately 600 m/s. The concepts of co- and counter-pellet-injection relative to the direction of rotation of the tokamak plasma are introduced. With an increase of the angle of pellet deviation, a substantial increase in the front duration and in the decay of bursts of radiation intensity of the deuterium Dα atom line was observed experimentally, and the simulations demonstrate an increase in the amplitude of the occurring radial electric field.

1. Introduction

Many phenomena in the high-temperature plasma, which is created and confined in tokamaks, are associated with the processes that take place at the edge of the plasma column near the last closed magnetic surface. Pellet injection is an active diagnostics, which substantially disturbs the plasma. It is one of the opportunities to study and change the confinement mode and transient processes due to the formation of strong gradients of density and temperature. Therefore, it is advisable to consider how to obtain the maximum density and temperature gradients, which preferably should be formed at a given radius of plasmas. It is necessary to choose the conditions and parameters both of the plasma and pellet injection to prevent the disruption of plasma discharge in the tokamak.

To achieve the maximum gradient, it is necessary to ensure long-term evaporation of pellets near magnetic surface of the selected target. At the edge, this can be done by reducing the velocity of injected pellets. Such experiments have been previously carried out [1, 2]. At a lower velocity, the pellet will evaporate at the edge of the plasma column and will not penetrate into the inner plasma regions. However, the reduction of pellet velocity decreases the stability of injection parameters and the precision of its synchronization with other systems that control processes in the tokamak plasma.

In this study, to solve the problem, instead of changing the speed of the particulates, it is proposed to change the direction of injection from radial to chord direction. In this case, the pellet will evaporate over longer-term near the magnetic surface that the injection direction is tangent to. Then, the maximum gradient forming region is pre-set not only at the edge of the plasma, but also in the interior of the plasma column. The minimum achievable radius for creating disturbances in this case can be calculated via the pellet evaporation scaling by taking into account changes in the plasma parameters along the chord direction of the injection. Compared with the central injection, for the chord injection, less pellet material will be deposited in the areas external to the area of maximum gradient, which increases the effectiveness of injection.

One of the tasks of this work was the technical implementation of the chord pellet injection and comparison of results of the chord injection with previous results of the central injection along the plasma radius [1–3].

Important parameters of the plasma at the last closed magnetic surface are the radial electric field, plasma rotation speed, and the shear. The moving pellet in the case of a central injection has no toroidal and poloidal velocity projections. During the evaporation and equalization of the background plasma temperature and particles in a cloud of evaporated pellet, a part of the energy of the toroidal and poloidal rotation transfers to the evaporated material. Due to this effect, the pellet "decelerates" plasma rotation differently at different radii. At the same time, the strong gradients give rise to the radial electric field, which significantly affects the shear of poloidal rotation [1].

In the case of chord pellet injection, one can distinguish between two types of injection geometry. In one case, pellet velocity is directed in the direction of plasma rotation, and the initial pellet impulse will provide a positive correction to the momentum of plasma rotation at the outer radii, and, in another case, the correction will be negative. The option of pellet injection in the same direction as the rotation of plasma we will name a co-pellet-injection, and injection in the direction opposite to the rotation of plasma - a counter-pellet-injection.

2. Experiment

For the experiments using co- and counter-pellet-injection, the chord injection system was developed at SPbPU, which allows to scan all directions from the central injection to injection at a tangent to the plasma column. The system was installed in the port that connects the pellet injector to the T-10 tokamak. The geometry of the non-central injection is shown in Figure 1. One of the objectives of this study was to show the possibility of deviation of accelerated deuterium pellets without their destruction. In addition, it was necessary to perform the deviation of pellets not at a fixed angle, but be able to smoothly change it from shot to shot, and thereby to scan a specific range of angles.

A modified multibarrel pellet injector (MPI-8, PELIN production) is used at the T-10 tokamak [4]. After modification, the MPI-8 device allows to inject up to five pellets of 1 mm in diameter and up to three - 0.7 mm in diameter per discharge with the range of pellet velocities of 500–800 m/s. Pellets are frozen inside the barrels according to the in situ technology. In this case, an 8-barrel injector is capable of injecting all pellets in a single discharge with an arbitrary delay.

The location of systems and diagnostics on the T-10 tokamak is shown in Figure 1. The fuel pellet injector is located in section "C" with a multi-channel microwave interferometer near the antenna of the diagnostics of the electron temperature, which monitors the change in intensity of the second harmonic of the electron cyclotron emission (ECE). Each section of the T-10 tokamak is equipped with gyrotrons for additional plasma heating. The SXR diagnostics is shifted by 90° in the toroidal direction relative to the pellet injection section. It is also possible to compare the results of deuterium injection and impurity pellet injection, which is performed in section "B", shifted by 90° relative to the deuterium injection device.

In May of 2016, the first experiments were carried out on the injection of up to 5 pellets in one discharge in regimes with additional plasma heating and using a chord pellet injection system. In addition to the central injection without pellet deviation, injection was performed with a deviation of 7.5° in the middle of a minor radius and 12.75° in the peripheral region of plasma. In these experiments, the direction of plasma current and toroidal field corresponds to the counter-clockwise rotation when viewing the T-10 tokamak from the top, as on the plan view in Figure 1. As the pellets are deflected up, for a given current direction, the experimental geometry corresponds to the counter-pellet-injection. Table 1 summarizes the main parameters of the injected pellets, which are considered in the article.

The experiments have shown that in the case of a deviation of the pellet injection angle from the central one, a fraction of destroyed pellets somewhat increases. However, in general, they have successfully passed through the chord pellet injection system and were deflected by a predetermined angle. During the plasma discharge, the chord injection system remains still, and the injection angle is fixed. During the time between the discharges, the angle of deflection can be changed to any given value of the operating range of the chord injection system.

Table 1. Parameters of the pellets that are injected at different deviation angles

Shot number	Deviation angle	Velocity v, m/s	Mass m, mg
68508	0°	671	0.35
68510	12.75°	685	0.36
68511	7.5°	662	0.38

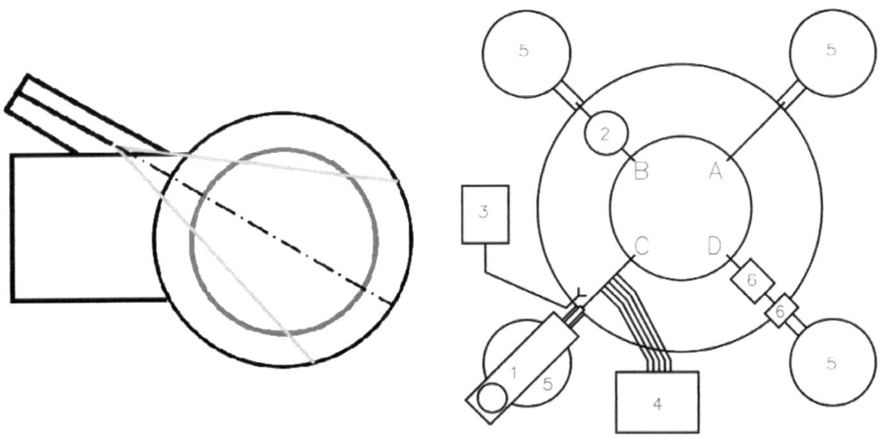

Figure 1. Scheme of the experiment on the T-10 tokamak.

The left drawing illustrates cross-section C with an installed chord pellet injection system. The dash-dot line corresponds to the case of central pellet injection. Solid green lines indicate a sector of available directions of the chord pellet injection.

The right scheme is a plan view of the T-10 tokamak. A, B, C, D are letters assigned to cross-sections with a set of ports in each of them, where 1 is the hydrogen pellet injector, 2 is the impurity pellet injector, 3 is the electron temperature diagnostics of plasma, 4 is the microwave interferometer, 5 are the gyrotrons, and 6 is the SXR diagnostics of plasma.

Figure 2 shows the plots of emission intensity of the Dα line of deuterium atoms for different angles of the counter-injection of deuterium pellets. Their comparison shows an increase in duration both of the rise time and the fall time of pulses with an increase of the angle of deviation of the injected pellets. For this dependence, there is still no definitive explanation and reproduction in the simulation results. Presumably, the observed delay in the formation of pulse Dα emission, in the case of deviation of injected pellets, is caused by longer evaporation time in the plasma peripheral layers, which are rapidly depleted in electrons with enough energy to evaporate the pellet material. This leads to a slower heating of the cloud of evaporated material and a stronger displacement of current to the central region of plasma, which considerably slows down the relaxation of peripheral regions after the injection.

IOP Conf. Series: Journal of Physics: Conf. Series **907** (2017) 012003 doi:10.1088/1742-6596/907/1/012003

3. Modeling

To analyze the experimental results, a comparative modeling of the central and chord injection should be performed. In the case of a chord injection, it is necessary to consider the two-dimensional geometry of the pellet trajectory and change the transport equation by incorporating the corrections, which are usually not taken into account when modeling the central injection.

To analyze the experimental results, modeling of the evolution of plasma was undertaken using the ASTRA code [5]. In this model, we considered a system of transport equations (1), where ρ is the effective minor radius, t is the time, V is the volume of plasma, $V' = \partial V/\partial \rho$, B_0 is the magnetic field at the center of plasma, Γ_e and Γ_i are the electron and ions fluxes, respectively; q_i and q_e are the corresponding heat fluxes, n_e and n_i are the density of electrons and ions, respectively; S_e, P_e and P_i are the terms that take into account the sources of electrons, electron and ion heat sources, respectively; $\sigma_{||}$ is the conductivity of the plasma in a direction parallel to the magnetic field; $J = I/R_0B_0$, where R_0 is the major radius of the tokamak vacuum chamber, I is the plasma current; $G = V'/(4\pi^2) \langle(\nabla \rho/r)^2\rangle$; j_{BS} and j_{CD} are the averaged density of bootstrap current and current drive, respectively.

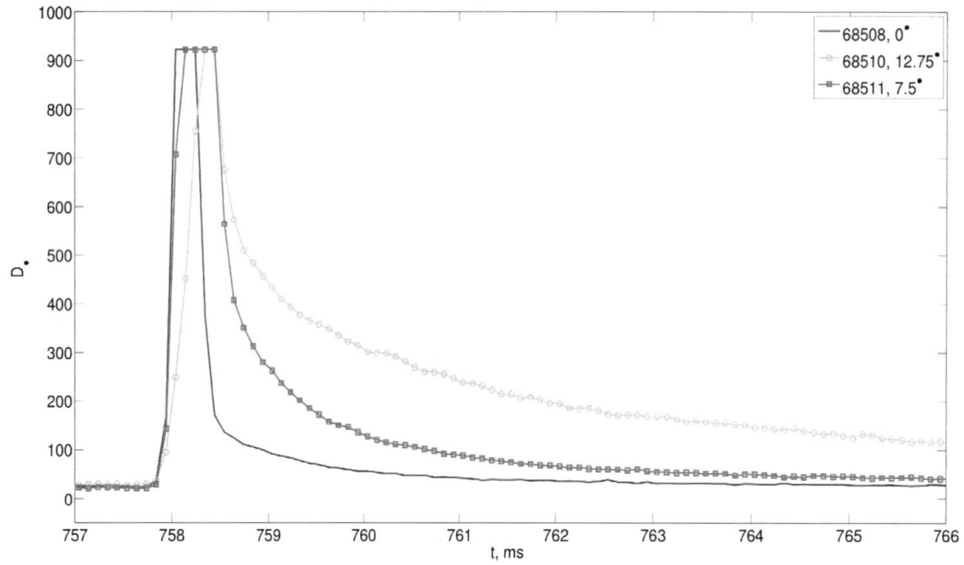

Figure 2. Spikes of the Dα line emission intensity after the pellet injection for 3 different pellet injection directions. Blue continuous line corresponds to 0° (central pellet injection), red line with squares - 7.5° deviation of the pellet trajectory, and green line with rhombs - 12.5° deviation (almost tangential injection).

$$
\begin{aligned}
&\frac{1}{V'}\left(\frac{\partial}{\partial t}-\frac{\dot{B}_0}{2B_0}\frac{\partial}{\partial \rho}\rho\right)(V'n_e)+\frac{1}{V'}\frac{\partial}{\partial \rho}\Gamma_e = S_e,\\
&\frac{3}{2}(V')^{-5/3}\left(\frac{\partial}{\partial t}-\frac{\dot{B}_0}{2B_0}\frac{\partial}{\partial \rho}\rho\right)\left[(V')^{-5/3}n_eT_e\right]+\frac{1}{V'}\frac{\partial}{\partial \rho}\left(q_e+\frac{5}{2}T_e\Gamma_e\right)=P_e,\\
&\frac{3}{2}(V')^{-5/3}\left(\frac{\partial}{\partial t}-\frac{\dot{B}_0}{2B_0}\frac{\partial}{\partial \rho}\rho\right)\left[(V')^{-5/3}n_iT_i\right]+\frac{1}{V'}\frac{\partial}{\partial \rho}\left(q_i+\frac{5}{2}T_i\Gamma_i\right)=P_i,\\
&\sigma_{\parallel}\left(\frac{\partial \psi}{\partial \rho}-\frac{\rho\dot{B}_0}{2B_0}\frac{\partial \psi}{\partial \rho}\right)=\frac{JR_0}{\mu_0\rho}\frac{\partial}{\partial \rho}\left(\frac{G_2}{J}\frac{\partial \psi}{\partial \rho}\right)-\frac{V'}{2\pi\rho}\left(j_{BS}+j_{CD}\right)
\end{aligned}
\qquad (1)
$$

The matrix of transport coefficients is given as (2). Expressions for the transport coefficients: α_i - neoclassical corrections that are small for the T-10 tokamak, r - minor radius, τ_h - on time of gyrotron, $\theta(\tau_h)$ - Heaviside function.

Figure 3 shows that the radial electric field E_r in case of injection without deviation is less than the simulation results for the maximum and average deviation. Since the evaporation takes place closer to the plasma edge for the injection with deviation, the radial electric field is also shifted to the periphery.

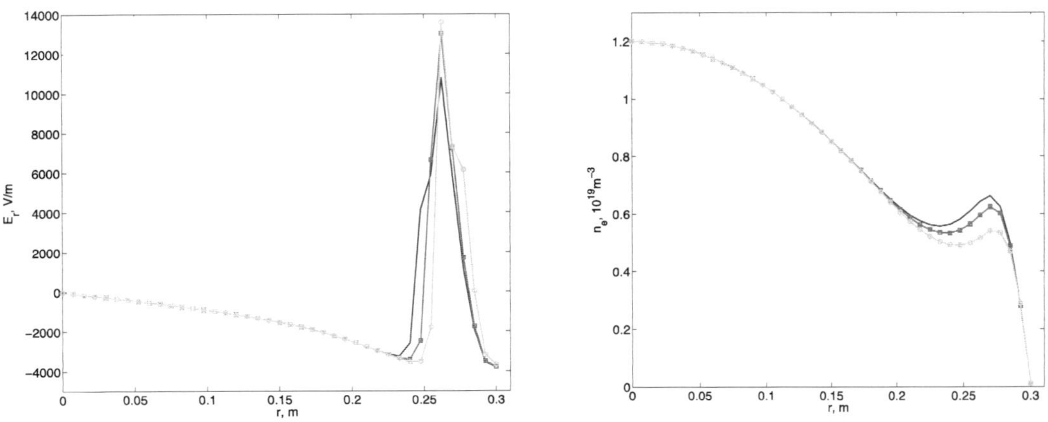

Figure 3. Results of modelling radial electric field (left plot) and electron density (right plot) in the case of chord pellet injections with different deviations from the axis of the central injection. Colors, symbols and lines are the same as in Figure 2.

$$
\begin{pmatrix} D_n & 0 & 0 & C_n \\ 0 & \chi_e & 0 & C_e \\ 0 & 0 & \chi_i & 0 \\ 0 & 0 & 0 & 0 \end{pmatrix}
\quad \text{where} \quad
\begin{aligned}
&D_n = 1+6\left(1-r^2\right)^{5/2}+\alpha_1\\
&\chi_e = 1+8\left(1-r^2\right)^4+2\theta(\tau_h)+\alpha_4\\
&\chi_i = 2+8\left(1-r^2\right)^5\\
&C_n=\alpha_2 \qquad C_e=\alpha_4
\end{aligned}
\qquad (2)
$$

4. Conclusions

The possibility of increasing density and temperature gradients of tokamak plasma as a result of chord injection of pellets has been experimentally demonstrated. It is shown that the change in the pellet injection angle makes it possible to create perturbations in the density and temperature of the plasma at the desired radius of the plasma. It was determined that in the case of chord pellet injection, the direction of rotation of the plasma should be taken into account. To denote two different variants of the combination of the direction of injection and rotation of the plasma, the terms "co-" and "counter-pellet-injection" are introduced, and the possibility of their application was realized. In the first case, the projection of pellet velocity on the plane, which is tangent to the plasma, coincides with the direction of rotation of the plasma, while in the second - they are opposite.

The chord pellet injection system was developed at the Peter the Great St. Petersburg Polytechnic University and installed on the T-10 tokamak. It allows to deviate accelerated pellets just before they hit the plasma. The system covers the entire range of angles from the center to the peripheral injection in both sides of the axis of the injector in a vertical cross section the T-10 tokamak. The first experiments have demonstrated the technical feasibility of a controlled deviation of the accelerated pellets without their substantial destruction at velocities of more than 600 m/s.

One of the observed differences in injection along different chords was an increase in the duration of the rise and fall times of intensity of the Dα line emission of deuterium atoms in comparison with central injection. Presumably, this is, firstly, due to greater depletion of the magnetic surface with primary electrons, which are spent evaporating the pellet material and, secondly, the displacement of plasma current in the central region of plasma.

The results of the experiments were modeled using the ASTRA code. The calculation of deposition of pellet material is performed by scaling for the evaporation rate [6] based on the actual weight of the pellet and taking into account the two-dimensional nature of its trajectory in the injection section. Simulations have shown the occurrence of a more intense radial electrical field in the case of chord injection compared with central injection.

Acknowledgements
The authors are grateful to I. V. Vinyar, P. V. Reznichenko, O. V. Ishevsky and N. M. Shepelev for the help with setting up the pellet injector. The authors thank the staff of the T-10 tokamak for providing the data and support. The work was carried out with the partial support from the State Corporation ROSATOM, contract No.H.4x.241.9B.17.1011.

References

[1] Kapralov V G, Rozhansky V A and Khlopenkov K V 1995 *Technical Physics Lett.* **21** 57
[2] Kapralov V G, Rozhansky V A and Khlopenkov K V 1995 *Proc. 22nd Eur. Conf. on Controlled Fusion and Plasma Physics (Bournemouth)* **19C** 1-117.
[3] Pavlov Yu D et al *21th IAEA Fusion Energy Conf. 2006 (China)* EX.P3-11 46
[4] Vinyar I V, Umov A P, Lukin A Ya and Reznichenko P V 2006 *Instruments and Experimental Techniques* **49** 717
[5] Pereversev G V and Yushmanov P N 2002 ASTRA - Automated System for TRansport Analysis *Preprint* IPP 5/98
[6] Kuteev B V 1999 *Technical Physics* **44** 1058 doi:10.1134/1.1259470

IOP Conf. Series: Journal of Physics: Conf. Series **907** (2017) 012004 doi:10.1088/1742-6596/907/1/012004

The regime of the improved confinement with deuterium pellet injected into plasmas of tokamak T-10 with W and Li limiters

D V Ryzhakov[1,3], Yu D Pavlov[1], A A Borschegovskiy[1], A V Gorshkov[1], V G Kapralov[2], L A Klyuchnikov[1], S V Krylov[1], S G Malzev[1], and D S Sergeev[1]

[1] NRC «Kurchatov Institute», Moscow, Russia
[2] Peter the Great St. Petersburg Polytechnic University, St. Petersburg, Russia

[3] E-mail: ryjakovdv@gmail.com

Abstract. In this paper, we present the first, after replacing a graphite limiter with a tungsten limiter, experimental results of the regimes of improved plasma confinement in the T-10 tokamak when injecting deuterium pellets. Comparison with the results of previous experiments with a graphite limiter shows the preservation of the improved confinement effect. Preliminary results of the experiments on the change in poloidal angle of injection of pellets allow us to say that with the central injection, the maximum effect of improved confinement is observed.

1. Experimental data and results

Studies of plasma scenarios on improved energy confinement and particle confinement during the injection of deuterium pellets in the T-10 tokamak have been performed for a long time [1]. The main purpose of these experiments is to study the regimes with improved plasma confinement. The T-10 tokamak limiter has a circular cross-section, main radius of 1.5 m, radius of unmovable circular limiter of 0.33 m, radius of movable limiter of 0.2–0.3 m, toroidal magnetic field $B_t \leq 5$ T. Before the experimental campaign of 2015–2016, the circular and movable limiters, which were made from graphite, were replaced with tungsten. As a result, it became possible to investigate the effect of impurities, their amount and distribution, on the characteristics of the regime of improved confinement when injecting deuterium pellets.

In the experiments, a modified multi-barrel pellet injector (MPI-8) from the Pelin firm was used. After the modification, MPI-8 allows to inject up to eight pellets per discharge (up to five pellets with a diameter of 1.0 mm and up to three pellets with a diameter of 0.7 mm) with a pellet velocity of 0.5–0.8 km/s.

In the analyzed shots with pellet injection, the plasma current was 270 kA, the toroidal magnetic field was 2.42 T, and the power of additional electron cyclotron heating was 0.85–1.5 MW. Similar to the earlier experiments with a graphite diaphragm [1–3], the injection of deuterium pellets during additional microwave heating in the experiments with a tungsten limiter, the plasma confinement was improved despite a much higher level of impurity content than with a graphite limiter. The change in the level of impurities after the replacement of graphite limiter with tungsten is described in detail in [4]. As can be seen from Table 1, when pellets are injected during electron cyclotron heating, the ratios of the experimental energy confinement time to the scaling ITER IPB98 (y, 2) are increased by 1.5 times, e.g., to increase the H_H factor.

$$\tau_{E,th}^{IPB(y,2)} = H_H 0.0562 I_p^{0.98} B_T^{0.15} P^{-0.69} n_{19}^{0.41} R^{1.97} M^{0.19} \varepsilon^{0.58} k_a^{0.78} \quad (1)$$

(in s, MA, T, MW, 10^{19} m^{-3} and m, respectively). Effective elongation is defined as $k_a = S_c / \pi a^2$, where S_c is the plasma cross-section area. The same H_H factor increase was also observed with a graphite limiter where internal and external transport barriers were observed [3]. The experimental energy confinement time was determined from diamagnetic measurements.

Table 1. Change of the H_H factor by scaling ITER IPB98 (y, 2) when pellets are injected for various limiter materials

	H_H factor before pellets during ECRH ($H_{H\ ECRH}$)	H_H factor after pellets ($H_{H\ pel}$)	$H_{H\ pel}$ / $H_{H\ ECRH}$
C limiter	0.7	≥ 1	≥ 1.5
W limiter	0.53	≥ 0.8	≥ 1.5

Improvement of the energy confinement is also demonstrated by other plasma characteristics. Thus, the loop voltage after pellet injection remained approximately at the same level as before the injection despite the density increase (see Figure 1). The vertical dashed lines in Figure 1 indicate the times of pellet injection.

Unlike the graphite limiter, with tungsten, we cannot confidently say about the presence of an internal transport barrier. However, in the ratio of the electron temperature profile after the injection of the first pellet to the temperature profile before the injection (Figure 2, 1-st pellet), there are regions with a high gradient. For all subsequent pellets, the corresponding profile ratio has a similar dependency (Figure 2, 2-nd pellet). This dependence is typical for internal transport barriers, the formation of which was demonstrated in experiments with a graphite limiter [1–3].

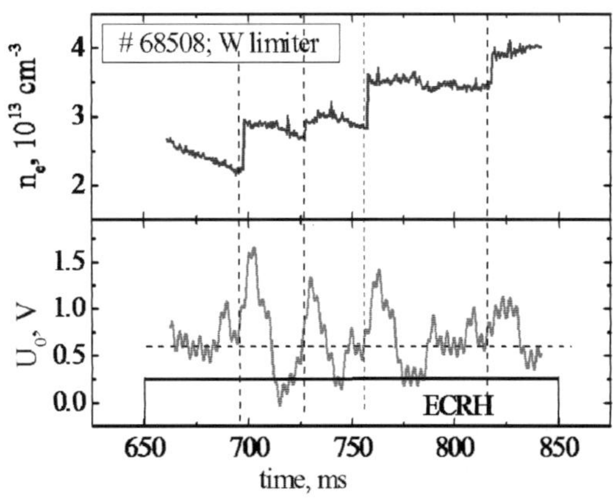

Figure 1. Average density along the central chord n_e and loop voltage U_0 at tungsten limiter.

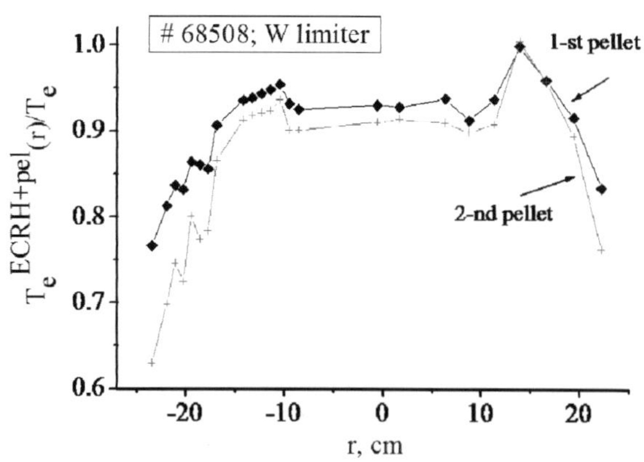

Figure 2. Ratio of the electron temperature profiles during the additional heating at tungsten limiter.

To check the efficiency of the chord system of pellet injection, several discharges with different poloidal injection angles were carried out. Schematically, the chord system of injection is shown in Figure 3. A preliminary analysis of these discharges makes it possible to say that the central injection leads to the greatest effect of improving confinement, because in that case, the H_H factor has the maximal increase (see Table 2).

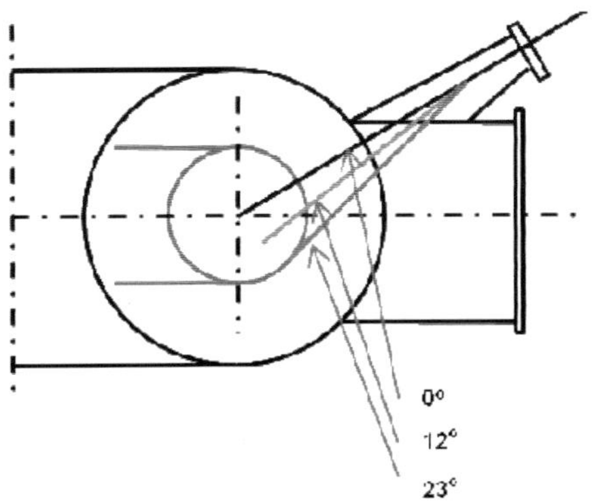

Figure 3. Pellet injection scheme at different angles.

Table 2. Change of the H_H factor by scaling ITER IPB98 (y, 2) with pellet injection at different angels

α	$H_{H\,pel}$	$H_{H\,ECRH}$	$H_{H\,pel}$ / $H_{H\,ECRH}$
0^0	0.8	0.53	1.5
12^0	0.72	0.53	1.3
23^0	0.62	0.53	1.1

2. Summary

Experiments with the conditions where the tokamak chamber is equipped with a tungsten limiter show that, despite the higher level of impurities than with a graphite limiter, the injection of pellets nevertheless leads to an improved confinement. Because the H_H factor with a graphite diaphragm increases by the same amount and transport barriers are observed, it can be assumed that even with a tungsten limiter, external transport barriers are formed.

Experiments to change the poloidal angle of pellet injection show that the greatest increase of the ratio of the experimental confinement time to the calculated value is observed with the central injection of pellets.

Acknowledgments.

The author would like to thank the staff of T-10 for participating in the experiment, and also I.V. Vinyar P.V. Reznichenko and O.V. Ishevsky for the engineering technical support of the pellet injector during the experiments. The work was supported by the Russian Atomic Energy Agency ROSATOM # H.4x.241.9Б.17.1011. The work on the chord injection system was carried out with the support of the grant RFBR # 14-02-00697-a.

References

[1] Pavlov Yu D et al 2000 *18th Fusion Energy Conference* IAEA-CN77/EXP/5/17, Sorrento, Italy

[2] Pavlov Yu D et al 2001 *28th EPS Conference on Controlled Fusion and Pl.Phys.* p4.020 Madera, Portugal

[3] Pavlov Yu D et al 2006 *21st Fusion Energy Conference, IAEA06/EX.P3-11* 46, China

[4] Krupin V A et al Experimental study of tungsten transport properties in T-10 plasma *Nucl. Fusion* **57** 066041

[5] ITER Phys. Expert Groups on Conf. and Transp. and Conf. Mod. and Database, ITER Physics Basis 1999 Editors // *Nucl. Fusion* **39** 2175–2249

Destruction of tungsten limiters in the T-10 tokamak under high plasma heat loads

S A Grashin[1,6], I I Arkhipov[2], V P Budaev[1,5], R N Giniyatulin[3], A V Karpov[1], L A Klyuchnikov[1], V A Krupin[1], N V Litunovskiy[3], I V Masul[3], F N Makhankov[3], Yu V Martynenko[1], D V Sarytchev[1], R Yu Solomatin[1] and L N Khimchenko[4]

[1] NRC "Kurchatov Institute", Moscow, Russia
[2] Institute of Physical Chemistry and Electrochemistry RAS, Moscow, Russia
[3] JSC "NIIEFA D. V. Efremov", Saint-Petersburg, Russia
[4] "Project center ITER", Moscow, Russia
[5] "National research university "MPEI", Moscow, Russia,

[6] E-mail: Grashin_SA@nrcki.ru

Abstract. Tungsten limiters were tested in the T-10 tokamak. The limiters were made from the ITER-grade WMP "POLEMA" tungsten. The influence of the edge tokamak plasma on tungsten limiters leads to significant cracking of tungsten. The heat load of up to 2 MW · m^{-2} leads to the micro-crack development at the grain boundaries accompanied by the loss of grains. The heat loads that exceed 5 MW · m^{-2} lead to the macro crack development. Under the present T-10 tokamak conditions, the heat and particle fluxes in the edge plasma lead to the significant destruction of tungsten limiters during the experimental campaign. During the disruption and runaway electron formation, extreme heat loads of more than 1 GW · m^{-2} cause strong melting of tungsten on the inner and outer part of the ring limiter.

1. Introduction

Tungsten is viewed as the plasma facing material for ITER [1, 2], as well as for the next generation fusion machines (such as DEMO). Thus, tungsten testing under high heat flux (HHF) loads during transient events (ELMs and disruptions) is very important for the ITER operation prediction. The investigation of tungsten melting and erosion under HHF was fulfilled on JET, AUG, T-10, LHD and a number of other fusion devices [1, 3, 4].

Under the real tokamak conditions, the combination of heat and particle fluxes can enhance tungsten damage and erosion. Different mechanisms in the edge plasma can concentrate energy on small zones of the divertor, limiters or wall. The heat loads, although less than the transient one (from ELMs, disruptions or runaway electrons), can be sufficiently high to cause damage or melting of tungsten. These effects should be examined under the real plasma condition to obtain valuable information for the ITER operation with tungsten. This paper presents the results of analysis of tungsten limiters of the T-10 tokamak after the 2015–2016 experimental campaign.

IOP Conf. Series: Journal of Physics: Conf. Series **907** (2017) 012005 doi:10.1088/1742-6596/907/1/012005

2. Experiments on the T-10 tokamak

During the T-10 operation with graphite limiters, significant cracking and erosion of graphite tiles was observed in the regions with the heat loads up to 1–2 MW · m^{-2}. The destruction of graphite tiles by the runaway electrons was local, but very intensive, which caused frequent (every 4–5 years) replacement of the tiles. In 2015, carbon limiters were replaced with tungsten limiters on the T-10 tokamak (Fig. 1). T-10 tokamak has a major radius R = 1.5 m, rail limiter with a radius of 0.30 m and a circular limiter of radius 0.33 m. The limiter was made at the "Efremov Institute" from the ITER-grade WMP "POLEMA" tungsten, which is used for the ITER divertor plates manufacturing. The impurities content does not exceed 0.05%, according to the certificate of the "Polema" company. Tungsten has high thermal conductivity and heat capacity. The thermal conductivity of VMP tungsten is 170–175 W/m K, which is 2.5 times higher than that of MPG-8 graphite. The heat capacity of the T-10 tungsten limiters is 2.5 times higher than of the limiters that are made from graphite.

Tungsten has a polycrystalline structure with a typical grain size of 30 μm and a porous surface. Cracks with a width of 1–2 μm were observed at the grains boundaries, as well as deeper in the material (Fig. 2). Cracks around the perimeter of grains can lead to chipping of the surface under high mechanical and thermal loads. Tungsten limiters of the T-10 tokamak do not have active cooling. The surface of the limiters was cooled via radiation and thermal conductivity.

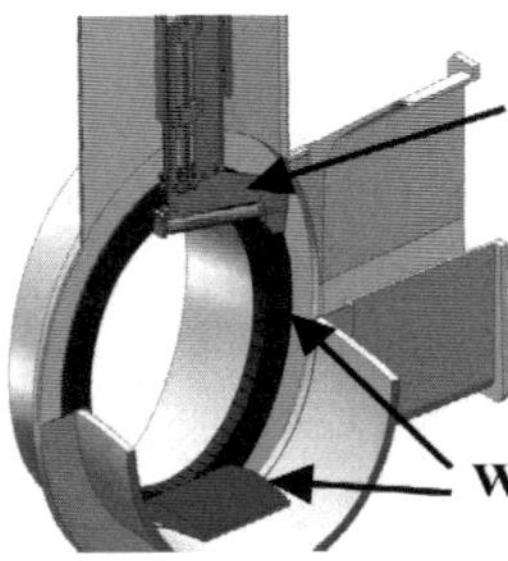

Figure 1. Scheme of the limiters in the T-10 tokamak.

Figure 2. Micrograph of the virgin surface of VMP "Polema" tungsten.

Heat flux from the plasma to the limiters was evaluated from the experimental data of Langmuir probes, which are located on the limiters and in the SOL of T-10. The temperature of the limiters was measured via thermocouples at a depth of 8–10 mm from the surface. The surface temperature of the limiters was evaluated from the visible radiation, which was recorded by fast optical camera.

3. Results

During the experimental campaign (November 2015 – December 2016), the T-10 tungsten limiters were exposed to ~2500 experimental shots. The normal T-10 operating shot duration was 1 s. Approximately 300–400 shots were terminated by disruptions and formation of the runaway electron beam. The ohmic heating power did not exceed 0.5 MW, and the ECR heating reached 2.5 MW during stable discharges, as well as during the discharges when disruption plasma interacts most intensely with the outer and inner part of the ring limiter (near the equatorial plane) and the top of the rail limiter (Fig. 3). Those zones suffered the strong destruction.

For rail limiter, the heat load does not exceed 1.5 MW · m^{-2} with ohmic heating and 2.5 MW · m^{-2} with ECR heating. The runaway electron beam is not deposited on the rail limiter surface. The surface temperature of the limiter estimated from the visible emission did not exceed 1500°C. The temperature on the bottom surface of tungsten tiles does not exceed 100°C. Strong development of the initial micro cracks was observed on the W-tiles under this conditions. The cracked structure became clear. In addition, loss of grains was observed on the tungsten surface after the experimental campaign (Fig. 4).

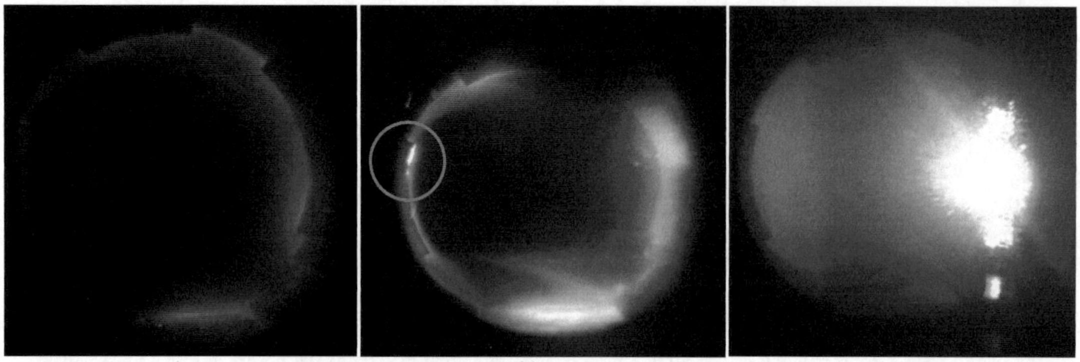

a) b) c)

Figure 3. Interaction of W limiters with the T-10 tokamak plasma recorded using a fast optical camera: (a) stable ohmic discharge; (b) discharge with ECR heating of 2 MW; (c) disruption and runaway electron beam formation.

Figure 4. Electron microscope images of the rail limiter W surface after the experimental campaign

On the inner and outer board of the torus, thermal loads on the edge of the limiter ring can reach 5 MW $\cdot$ m^{-2} during ohmic heating, and exceed this value during ECR heating. For the case shown in Fig. 3b, the temperature of the ring limiter surface exceeds 1500°C. Although the mechanism of local energy release on the limiter surface has not been identified yet, it should be noted that such heating cannot be explained by the action of runaway electrons. Cracking in these zones leads to the formation of macro cracks with a width up to 80 microns and lengths from 1 to 50 mm. Cracks on the surface of tungsten plates are directed in longitudinal and transverse directions, forming a net of cracks (Fig. 5). The edges of some cracks are melted. The micro crack structure at the grain boundaries is also observed on the rail limiter W-surface. In the areas with the highest heat loads, tungsten plates are melted and are close to complete destruction.

At the runaway electron beam contact area (at the inner and outer part of the ring limiter near the tokamak equatorial plane), thermal loads of more than 1 GW/m^2 cause strong melting of the tungsten surface (Fig. 6) [4]. In addition, the melted region suffered significant cracking. Droplets of the melted tungsten with sizes up to 2–3 mm and redeposited tungsten are only observed in the vicinity of the runaway electron strike zone. No melted or redeposited W was observed in other parts of the T-10 vacuum vessel.

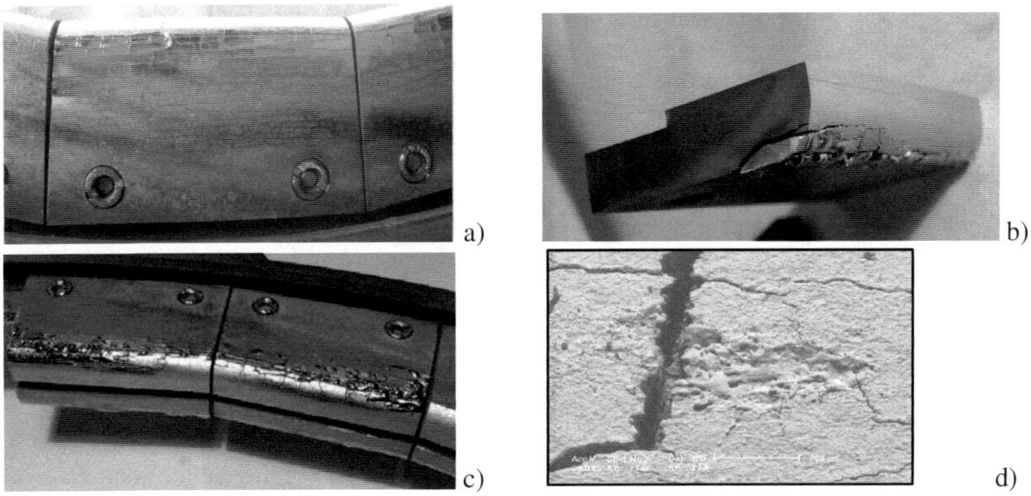

Figure 5. Tungsten plates from the inner side of the ring limiter after the experimental campaign (a–c). Melting and cracking of the tungsten surface is observed. Electron microscope micrograph of cracks on the tungsten surface (d).

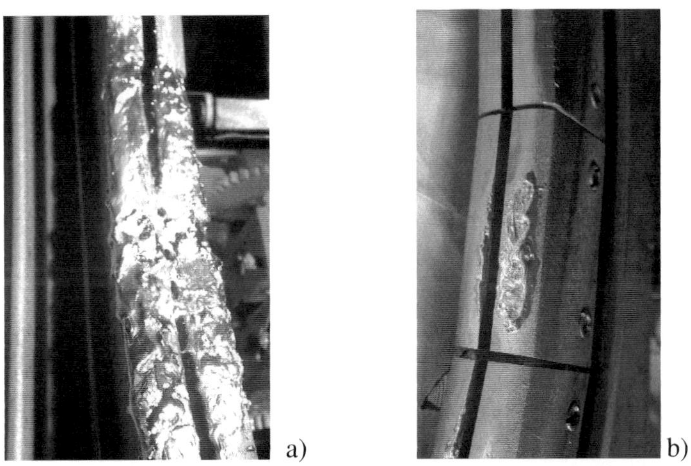

Figure 6. Melting of tungsten plates on the (a) outer and (b) inner boards of the ring limiter at the runaway electron beam contact area.

4. Conclusions

In 2015, carbon limiters were replaced with tungsten limiters on the T-10 tokamak. The limiters were made from the ITER-grade WMP "POLEMA" tungsten, which is used for the divertor plate manufacturing. WMP tungsten does not contain any impurities and has high thermal conductivity and heat capacity. Tungsten has a polycrystalline structure and a porous surface.

The influence of the edge tokamak plasma on tungsten limiters leads to the significant cracking of tungsten. The heat load up to 2 MW · m^{-2} leads to the micro-crack (1–2 μm wide) development at the grain boundaries accompanied by the gradual destruction of the surface including loss of grains. Heat loads exceeding 5 MW · m^{-2} lead to the macro crack development up to an 80-μm width and a length of 5–50 mm. There is no correlation of the development of such cracks with runaway electrons. Under the present T-10 tokamak conditions, the combination of heat and particle fluxes in the edge plasma lead to the strong destruction of tungsten limiters during the experimental campaign. With an expected rise of the ECRH power up to 3.5 MW, the limiter damage should be much stronger

During the disruption and runaway electron formation, extreme heat loads of more than 1 GW/m^2 caused strong melting of the tungsten plates on the inner and outer part of the ring limiter near the equatorial plane of the tokamak. Near this zone, the droplets of molten tungsten and redeposited tungsten layers were observed. In other areas of the tokamak chamber, no layers or drops of eroded tungsten were detected.

Acknowledgments
The authors express their gratitude to the T-10 staff for assistance with the experiments. The work is supported by ROSATOM (Contract H. 4.241.9 B. 17.1011).

References

[1] V P Budaev 2016 *Physics of Atomic Nu*clei **79**(7) 1137-1162
[2] R A Pitts, S Carpentier, F Escourbiac et al 2013 *Journal of Nuclear Materials* **438** 48
[3] V P Budaev, Yu V Martynenko, et al 2015 *Journal of Nuclear Materials* **463** 237–240
[4] V P Budaev, Yu V Martynenko, S A Grashin et al 2017 *Nuclear Materials and Energy*

IOP Conf. Series: Journal of Physics: Conf. Series **907** (2017) 012006 doi:10.1088/1742-6596/907/1/012006

Plasma arcs formation in the plasma periphery during disruptions in the T-10 tokamak plasma

P V Savrukhin[1], E A Shestakov, A I Ermolaeva and R Yu Solomatin

National Research Center "Kurchatov Institute", Moscow

[1] E-mail: psavrukhin@bk.ru

Abstract. Fast-scale plasma perturbations (0.5–2 MHz) are studied during disruption instability in the T-10 tokamak. A system of movable magnetic probes is used to measure magnetic field perturbations in the plasma periphery. Spatial distribution of soft X-ray oscillations is measured using silicon surface-barrier detectors. A possible connection between the fast-scale electromagnetic and X-ray oscillations and arc discharges in the plasma edge near localization of the MHD modes during plasma disruption is considered.

1. Introduction

Stabilization of disruptions and prevention of accelerated electron beams are important tasks in the design of a reliable thermonuclear reactor. Practical implementation of disruption stabilization systems is complicated due to the variety of plasma perturbations, which are observed prior to the plasma collapse. The initial stage of disruption is associated, as a rule, with the nonlinear growth of magnetohydrodynamic (MHD) modes with low poloidal and toroidal numbers ($m = 1–5$, $n = 1–3$). However, the amplitude of MHD perturbations can differ by an order of magnitude before the disruptions under various experimental conditions, which leads to the need for complex adaptation of the control system parameters. The critical conditions of disruption can be detected using multiparametric analysis with neural networks. However, the adjustment of such systems requires a preliminary study of a significant number of disruptions, which is difficult under the tokamak reactor conditions. Analysis of the critical conditions for the energy disruption development and the generation of trigger signals for the initiation of disruption stabilization systems is one of the main tasks of modern experiments.

Experiments on the T-10 tokamak have shown that the initial stage of plasma disruptions at high density is accompanied by the development of rapidly varying oscillations of electromagnetic fields [1], which are possibly associated with the formation of arc plasma discharges in the peripheral regions of plasma. The formation of local arc discharges, which has been studied in detail earlier in the laboratory experiments with a low-temperature plasma, is indirectly confirmed in the T-10 tokamak plasma in the form of characteristic damages of the tokamak vessel elements. Registration of high-frequency perturbations of magnetic fields at the initial stage of disruption can be used to generate trigger signals for the initiation of disruption stabilization systems.

2. Diagnostic systems of magnetic probes and soft X-ray radiation in the T-10 tokamak

Magnetic probes are widely used in the studies of plasma perturbations in tokamak research. In the experiments with high-temperature plasma, magnetic probes are usually located at a considerable distance from the plasma boundary and are protected by additional screens of graphite and stainless steel. Such an arrangement of probes makes it difficult to measure fast-oscillating plasma oscillations with frequencies above f ~ 100 kHz. An additional limitation is associated with currents that are induced in conductive structures on which the magnetic probes are fixed.

The system of magnetic probes in the T-10 tokamak [1] consists of the horizontal and vertical coils that are fixed to a movable rod. The probes are made of nickel wire in ceramic insulation and are located inside a ceramic container that is separated from a supporting metal structure. The signals from magnetic probes are digitized via ADC converters ADM214x100M (14 bits, 60 MHz).

To study the fast-scale soft X-ray oscillations, a diagnostic tool that is based on silicon surface-barrier detectors DKPs-50 with a high-speed amplifier is installed in the T-10 tokamak. The X-ray data acquisition system is based on an L-783M multifunction high-speed ADC board (12 bits, 3 MHz) that is located in close proximity to the T-10 tokamak.

3. Results of the experiments in the T-10 tokamak

A system of magnetic probes was used in the experiments in the T-10 tokamak to study plasma perturbations during the density limit disruption, which is induced by additional gas puffing at the quasi-stationary stage of the discharge. The typical evolution of plasma parameters in these experiments is shown in Fig. 1a. A thermal quench (energy collapse) is initiated at time t ~ 719.4 ms. Along with large-scale MHD perturbations (m = 2–3, n = 1), magnetic fluctuations in the fast-scale range (f_{fast} ~ 0.2–0.5 MHz) are observed just before the thermal quench. Amplitude of the fast-scale oscillations decreases much more rapidly than the amplitude of the m = 2 mode when the probe is moved away from the plasma boundary (see Fig. 1b). Analysis shows that the fast-scale oscillations are characterized by a wide spectrum of harmonics (see t = 719.45–719.51 ms in Fig. 2a).

To reveal the dominant perturbations of magnetic signals, a wavelet transformation is used in the analysis. An example of the analysis of the magnetic probe signal using continuous wavelet transformation (CWT) algorithms, which are based on Daubechies wavelets of the 4th order, is shown in Fig. 2b. This plot confirms the multi-scale origin of perturbations and highlights periodicity, which is observed at a scale coefficient of the 50th order (which is equivalent to regular oscillations at frequencies f < 450 kHz). Inspection of the CWT coefficients plot for this signal reveals patterns among the scales that show some features of the signal's chirping from high (low frequencies) to low (high frequencies) scales (see, for example, t > 719.49–719.51 ms in Fig. 2b).

At the initial stage of disruption, fast-scale perturbations are also observed in the form of soft X-ray oscillations (f ~ 0.5–1 MHz), which are recorded using Si detectors with a direct view of the graphite diaphragm that limits the plasma boundary (see Fig. 3). This indicates that there may be a possible connection between the fast-scale perturbations and processes of interaction of local plasma discharges with the surface of limiters.

When the large-scale MHD perturbations are rotated, the fast-scale oscillations manifest themselves most clearly at the moment of passage of the O-points of the magnetic islands near magnetic probes, which are located at the boundary of the annular poloidal limiter (see Figure 4). This effect can be related to the enhancement of the plasma-wall interaction during the passage of the O-points of the magnetic islands near the surface of the limiter. The role of the interaction of plasma with the limiter surface in the development of fast-scale oscillations is also confirmed by a significant decrease (up to 100 times) of the amplitude of magnetic perturbations, which is measured with magnetic probes that are located far from the limiters (when displaced in the toroidal direction by 180 degrees).

4. Discussion and Conclusions

The new system of magnetic probes is installed in the T-10 tokamak. The system is based on high-sensitivity magnetic probes that are located inside the vacuum vessel on a movable rod, which ensures the positioning of probes near the plasma boundary. The important result of this study is the detection of rapidly varying (0.2–0.5 MHz) perturbations of magnetic fields at the initial stage of disruption. Measurements of the radial and angular distributions of the toroidal and poloidal components of the fast-scale magnetic perturbations show that the fast-scale perturbations cannot be connected with the helical perturbations (micro-tiring mode). The detection of similar fast-scale perturbations of soft X-ray radiation in the plasma periphery indicates a possible connection between the fast-changing oscillations and the processes near and on the surface of diaphragms that limit the plasma discharge boundary.

The analysis of damage to the limiters in the T-10 tokamak confirms the development of arc plasma discharges in the peripheral regions of plasma. In addition, perturbations in the range of a few MHz were observed during arc discharges in laboratory experiments with an electric spark (50 V, 120 J), which was used for calibrating the magnetic probes in the T-10 tokamak.

Experiments, which were carried out on the T-10 tokamak, have shown a possible connection between the fast-scale magnetic and X-ray oscillations and arc plasma discharges in the peripheral regions of the discharge. Registration of the high-frequency magnetic field perturbation at the initial stage of disruption can be used to generate trigger signals for switching on the systems of disruption stabilization.

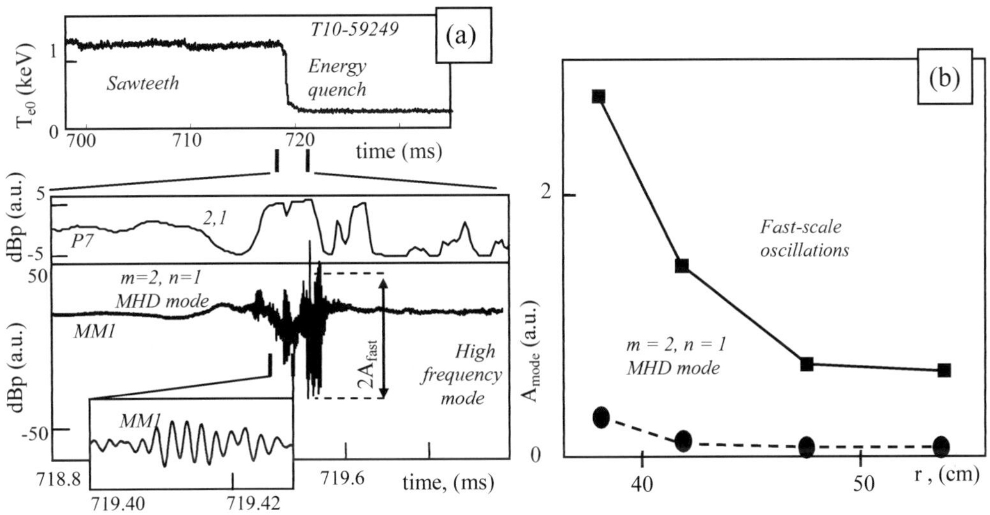

Figure 1. (a) Temporal evolution of plasma parameters in the T-10 tokamak in the regime with disruption at high density. Along with large-scale MHD perturbations (m = 2, n = 1), fluctuations in the range f_{fast} ~ 0.2–0.5 MHz are observed just before the thermal quench (t ~ 719.4 ms). (b) Radial dependence of the oscillation amplitudes due to the m = 2 mode (circles) and fast-scale oscillations of magnetic fields (squares).

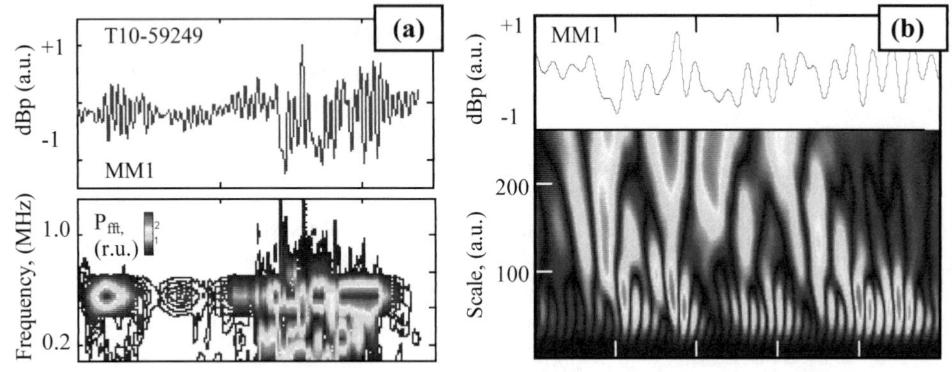

Figure 2. Time evolution of magnetic field perturbations in the T-10 density limit disruption. Also shown are the results of (a) Fourier analysis of magnetic perturbations and (b) the image of expansion coefficients in wavelet functions.

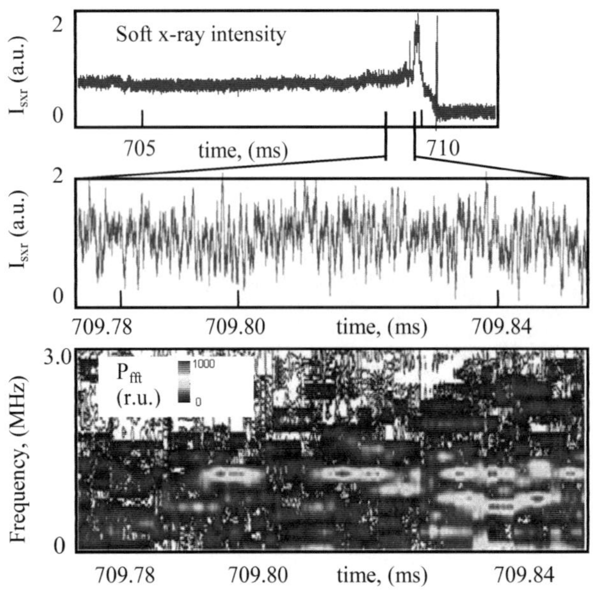

Figure 3. Time evolution of the soft X-ray intensity and results of the Fourier analysis of fast-scale X-ray perturbations in the T-10 density limit disruption.

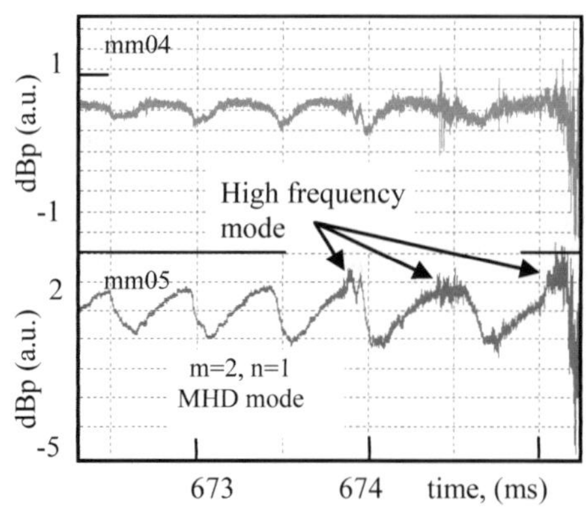

Figure 4. Time evolution of magnetic field perturbations prior to the density limit disruption. Fast-scale magnetic perturbations are observed most clearly close to the O-points of magnetic islands.

Acknowledgments

The paper is partly supported by Rosatom H.4x.241.9Б.17.1011 and the NRC "Kurchatov Institute" order 625.

References

[1] Savrukhin P V and Shestakov E A 2012 *Rev. Sci. Instrum.* 83 013505 http://dx.doi.org/10.1063/1.3675577509.

IOP Conf. Series: Journal of Physics: Conf. Series **907** (2017) 012007 doi:10.1088/1742-6596/907/1/012007

Effect of ECRH and resonant magnetic fields on formation of magnetic islands in the T-10 tokamak plasma

E A Shestakov[1] and P V Savrukhin

National Research Center "Kurchatov Institute", Moscow

[1] E-mail: shestakovj@mail.ru

Abstract. Experiments in the T-10 tokamak demonstrated possibility of controlling the plasma current during disruption instability using the electron cyclotron resonance heating (ECRH) and the controlled operation of the ohmic current-holding system. Quasistable plasma discharge with repeating sawtooth oscillations can be restored after energy quench using auxiliary ECRH power when $P_{EC} / P_{OH} > 2–5$. The external magnetic field generation system consisted of eight saddle coils that were arranged symmetrically relative to the equatorial plane of the torus outside of the vacuum vessel of the T-10 tokamak to study the possible resonant magnetic field effects on the rotation frequency of magnetic islands. The saddle coils power supply system is based on four thyristor converters with a total power of 300 kW. The power supply control system is based on Siemens S7 controllers. As shown by preliminary experiments, the interaction efficiency of external magnetic fields with plasma depends on the plasma magnetic configuration. Optimal conditions for slowing the rotation of magnetic islands were determined. Additionally, the direction of the error magnetic field in the T-10 tokamak was determined, and the threshold value of the external magnetic field was determined.

1. Introduction

One of the most dangerous consequences of plasma disruption is the formation of runaway electron beams [1]. Analysis showed that runaway electrons can gain high energy (up to 30 MeV) and release it locally in the form of heat by interacting with the vacuum vessel internal elements. Moreover, high power density of the runaway beams (up to 3 GW/m^2) [2] leads to surface damage or total destruction and failure of the vacuum vessel internal elements.

In previous experiments on tokamaks, it was shown that large magnetic field perturbations, in particular low-frequency and stationary MHD modes, can lead to weakening and disappearance of the runaway electron beams [3]. Therefore, suppression of the runaway electrons can be provided by reducing the longitudinal electric fields via ohmic and auxiliary plasma heating system and an MHD perturbations control system.

2. Effect of ECRH on dynamics of the MHD perturbations and runaway electron beams

Effect of ECRH on MHD perturbations and runaway electrons was studied in the T-10 tokamak in experiments with discharge recovery after density limit disruption [4]. The experiments showed that auxiliary ECRH is a flexible tool for controlling MHD perturbations in plasma, which allows to both stabilize the MHD perturbations and create large-scale MHD modes.

The results of experiments on the study of the ECRH effect on the dynamics of MHD perturbations and runaway electrons after thermal quench at limiting density are shown in figure 1. Energy quench

IOP Conf. Series: Journal of Physics: Conf. Series **907** (2017) 012007 doi:10.1088/1742-6596/907/1/012007

is usually accompanied by a series of total radiation power losses and hard X-ray radiation bursts that appear during the interaction of runaway electrons with edge plasma and limiter. Deposition of the ECR power in the central plasma regions (figure 1a, B_t = 2.4 T) leads to the formation of quasi-stationary MHD perturbations and subsequent disappearance of radiation bursts as well as runaway electrons.

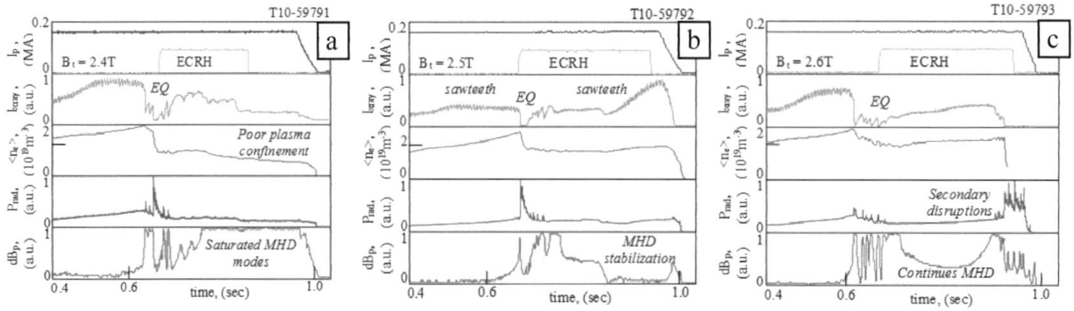

Figure 1. Plasma parameters during density limit disruption on the T-10 tokamak with different toroidal magnetic field B_t amplitudes. I_p is the plasma current; I_{xray} is the soft X-ray radiation intensity (gas detector); $<n_e>$ is the average electron density; P_{rad} is the total radiation power loss (bolometer); and dB_p is the time derivative of the amplitude of magnetic perturbations (magnetic coils).

Deposition of the ECRH power near the q = 2 resonant surface (figure 1b, B_t = 2.5 T) leads to stabilization of the MHD perturbations and complete disappearance of the runaway electron beams. Deposition of the ECR power outside the q = 2 resonance surface (figure 1c, B_t = 2.6 T) leads to the partial stabilization of MHD perturbation and disappearance of the nonthermal radiation bursts. However, after switching off the ECRH, the plasma becomes unstable and disruption occurs with a subsequent appearance of the secondary runaway electron beams.

3. Stationary MHD perturbations generation system on the T-10 tokamak

To study the effect of resonant magnetic fields on plasma stability, the T-10 tokamak was equipped with a system of eight saddle coils that are arranged symmetrically relative to the equatorial plane of the torus outside the tokamak vacuum vessel (figure 2). This geometric configuration allows to generate quasi-stationary magnetic fields with a toroidal harmonics n = 1–7. The windings have a rectangular shape, their height is 1.5 m, and length along the torus is 2 m. The windings have eleven turns and are wound with a cable with a conductor cross section of 120 mm^2.

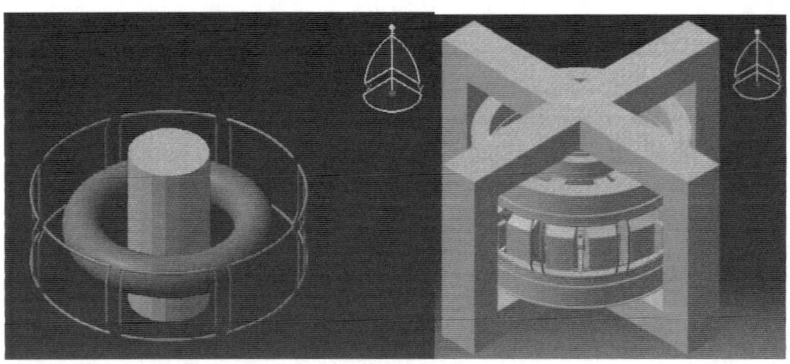

Figure 2. Schematic view of external saddle coils on the T-10 tokamak

Four independent power supplies with a total power of 300 kW are used to power the saddle coils. The type of power supply used is a thyristor converter with a rated current of 1250 A and the possibility of short time excess up to 1900 A. The maximum output voltage is 45 V. The power supply can be remotely switched on/off, and the output voltage can be regulated. The amplitude of external magnetic perturbations m = 2, n = 1 near the q = 2 resonant surface is approximately 20–30 G.

The saddle coils power supply control system, which is based on a programmable logic controller Siemens S7-417, was designed and equipped on the T-10 tokamak. The control system measures basic parameters of power sources (current, voltage, and logic current sensor), determines their operating mode, and records parameters in a database. The data is transmitted over the Profibus network and recorded on a computer in a common local area network of the T-10 tokamak control room.

4. Effect of resonant MHD perturbations on the magnetic islands in tokamak T-10 plasma

Experiments were carried out on the T-10 tokamak discharges with various plasma parameters (I_p = 120–290 kA, B_t = 1.5–2.4 T). In these experiments, the total magnetic field of the windings with one pair of magnetic poles was created, which corresponds to the n = 1 magnetic configuration. Figure 3 shows the evolution of MHD perturbations in experiments with saddle coils operating in the optimal regime for controlling the MHD perturbations.

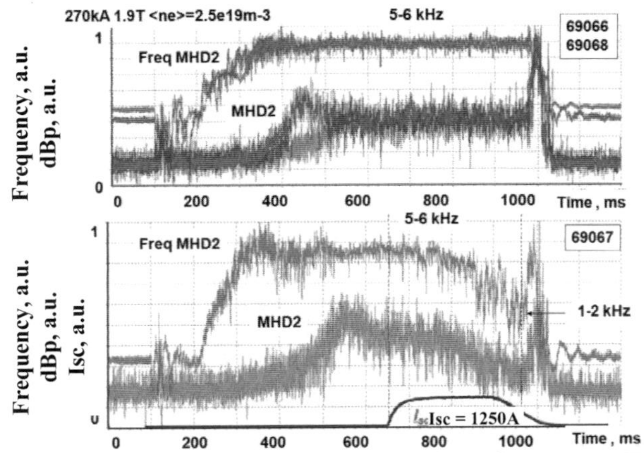

Figure 3. Time evolution of the MHD perturbations frequency (Freq) and amplitude (MHD2) in experiments with external magnetic fields (Isc is the amplitude of the total current in the saddle coils).

Typically, at a sufficiently large amplitude of external perturbation (figure 3, pulse 69067), the switching on of saddle coils slowed down the m = 2 magnetic perturbation rotation speed from 5 kHz to 1–2 kHz. The strongest effect on slowing the rotation of MHD modes is observed when the q = 3 magnetic surface is placed close to the plasma boundary (figure 4).

Preliminary analysis indicated that the direction of magnetic field at which the MHD mode rotation stops most effectively coincides with the direction of the tokamak T-10 error field. It is also shown that the rate of MHD perturbations slowing increases nonlinearly at a threshold current in saddle coils (figure 5).

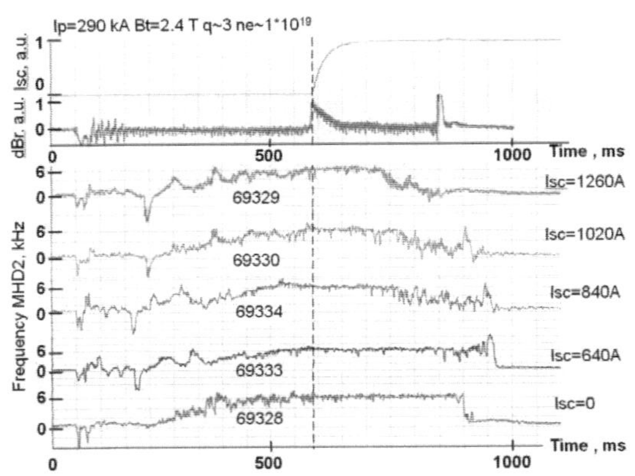

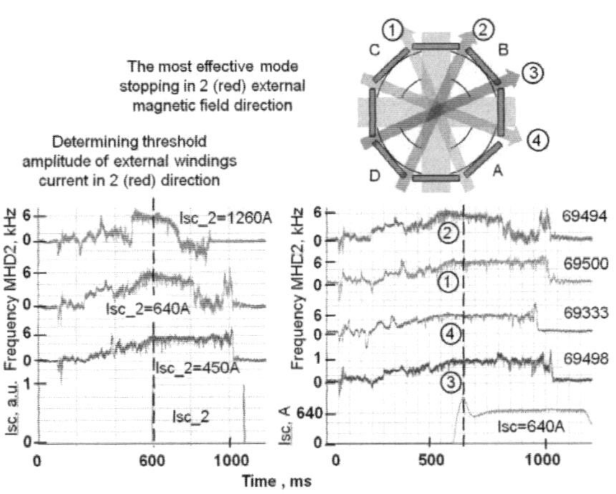

Figure 4. Mode m = 2 rotation slowing at various currents I_{sc} in the external coils (I_p = 290 kA, Bt = 2.4 T, $n_e \sim 1 \cdot 10^{19}\ m^{-3}$)

Figure 5. Experimental results of the T-10 tokamak error filed direction and external magnetic field threshold value determination

5. Conclusions

Effects of the ECRH auxiliary heating and resonant magnetic fields on plasma stability were studied in the T-10 tokamak. The possibility of safe recovery of stable plasma discharge and runaway electron beam stabilization was shown when using ECRH for plasma heating and MHD perturbation burst stabilization. The external resonant magnetic fields generating system based on eight saddle coils is installed in the T-10 tokamak. The system is powered by thyristor converters and controlled via Siemens S7 controllers. Experiments with saddle coils demonstrated nonlinear effects of external magnetic fields on the m = 2 mode rotation speed. Preliminary analysis indicated that the effect of external magnetic fields on plasma stability strongly depends on the position of internal resonant magnetic surfaces with respect to the saddle coils and on the angular direction of external magnetic fields.

Acknowledgements

This work was supported by ROSATOM State Corporation (Contract H.4x.241.9Б.17.1011) and the NRC “Kurchatov Institute” order 625.

References

[1] Savrukhin P V and Shestakov E A 2015 *Nucl. Fusion* **55** 043016 9
[2] Budaev V P et al 2016 *Nucl. Materials and Energy* http://dx.doi.org/10.1016/j.nme.2016.11.029
[3] Yoshino R and Tokuda S 2000 *Nuclear Fusion* **40** 1293-1309
[4] Savrukhin P V and Shestakov E A 2016 *Physics of Plasmas* **23** 112509

IOP Conf. Series: Journal of Physics: Conf. Series **907** (2017) 012008 doi:10.1088/1742-6596/907/1/012008

Collisionless distribution function of charged particles ensemble in a tokamak magnetic configuration with magnetic island

O I Podturova[1,2,3]

[1] RUDN University, 6 Miklukho-Maklaya Street, Moscow, 117198, Russia
[2] National Research Center "Kurchatov Institute", 1 Akademika Kurchatova pl., Moscow, 123182, Russia

[3] E-mail: olga_podturova@list.ru

Abstract. The collisionless distribution function of charged particle ensemble in the magnetic field of tokamak with a magnetic island is calculated. The calculation is based on the solution of the kinetic equation with source together with three-dimensional numerical calculations of charged particle trajectories. It is shown that in case of an inhomogeneous source trajectory, motion of trapped particles leads to anisotropization of the initially isotropic distribution of particle ensemble. The absence of contribution from the passing particles decreases the efficiency of spontaneous generation of a non-induction current in the magnetic island in comparison with the bootstrap effect in the system of nested magnetic surfaces.

1. Introduction

It is well known that the motion of charged particles, whose initial distribution is not a function of integrals of motion, leads to evolution of their distribution function and its spontaneous anisotropization. In particular, the consequence of this anisotropization in a tokamak toroidal-helical magnetic field is the effect of bootstrap current generation [1, 2], which has the key value for the non-inductive current drive in a stationary tokamak reactor [3]. The maximum possible bootstrap current density, which is generated due to the spontaneous anisotropization of charged particle ensemble as a result of their interaction with the axisymmetric tokamak magnetic field, was calculated in [4, 5] by solving the collisionless kinetic equation. Meanwhile, in modern tokamaks, three-dimensional effects play a crucial role. The appearance of magnetic islands under the influence of external resonant magnetic perturbations and/or the development of tearing instabilities drastically change plasma equilibrium and transport. The theory of related processes is poorly developed, primarily due to the absence of symmetry in the problem. In this paper, using the method proposed in [4], we calculate the collisionless distribution function in the center of magnetic island and make qualitative conclusions about the efficiency of bootstrap current generation in such system.

2. Calculation Method

We represent magnetic field in terms of the Hamiltonian function [6], which guarantees the fulfillment of the solenoidal condition for vector **B,** by the construction: $2\pi\mathbf{B} = [\nabla\Phi \times \nabla\theta] + [\nabla\Psi \times \nabla\varphi]$. Here, θ is the poloidal angle, φ is the toroidal angle, Φ is the toroidal magnetic flux function, and Ψ is the poloidal magnetic flux function. The safety factor in this formulation is given by the expression:

$q = -d\Phi / d\Psi$. To determine the concrete magnetic configuration, we define Φ and Ψ as functions of coordinates {ρ, φ, θ} of a quasi-cylindrical system that is related to the tokamak magnetic axis. The value ρ = 0 corresponds to the magnetic axis, ρ = 1 corresponds to the plasma boundary. The function Φ is chosen in such a way that the toroidal magnetic field decreases proportionally to $1/r$; r is the distance from the geometrical axis of the torus, $r = R + \rho\cos\theta$ (R is the major tokamak radius): $\Phi = \rho^2(1 - 2\rho\cos\theta / 3R)/2$. The poloidal magnetic flux function Ψ can be written in the following form:

$$\Psi = -\frac{\Phi}{1+2\Phi} + \Phi(1-2\Phi)A_{3/2}\cos(2\varphi - 3\theta). \tag{1}$$

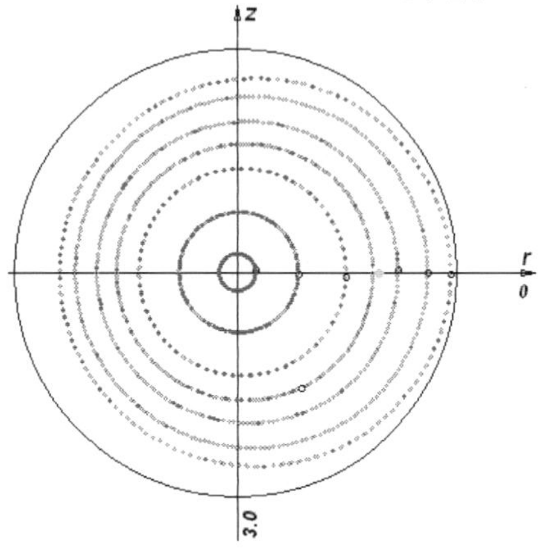

Figure 1. Poincaré cross-section of the unperturbed magnetic configuration, $A_{3/2} = 0$.

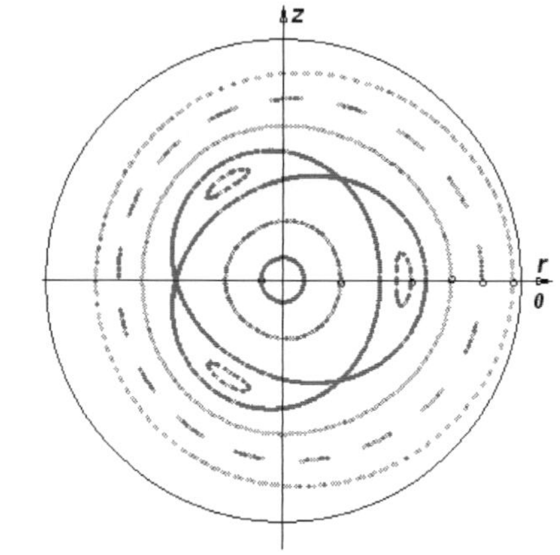

Figure 2. Poincaré cross-section of the magnetic configuration with an island, $A_{3/2} = 0.01$.

The first term in the right-hand side of expression (1), $\Psi_0 = -\Phi/(1+2\Phi)$, sets an unperturbed system of nested magnetic surfaces that are symmetric with respect to the equatorial plane of the tokamak, see figure 1, with the profile of safety factor $q = (1+2\Phi)^2$ (q varies from 1 on the magnetic axis to 4 at the plasma boundary). The second term on the right-hand side of expression (1) describes the perturbed part of the Hamiltonian Ψ: the coefficient $A_{3/2}$ determines the amplitude of helical perturbation; m = 3 and n = 2 are the numbers of poloidal and toroidal harmonics, respectively; the factor Φ (1 - Φ) ensures that the perturbation in the center and at the plasma boundary is zero. The appearance of a resonant helical perturbation ($A_{3/2} \neq 0$) leads to the splitting of the rational surface q = 3/2 with the formation of chain of islands in the middle of the plasma column, see figure 2.

To calculate the collisionless distribution function, f, we use the method that has been proposed and implemented in a series of works [4, 5, 7]. We describe the evolution of distribution function of charged particle ensemble in a magnetic by the collisionless kinetic equation with source S:

$$\frac{df}{dt} = \frac{\partial f}{\partial t} + \mathbf{v}\nabla f + \frac{Ze}{m}\frac{[\mathbf{v}\times\mathbf{B}]}{c}\frac{\partial f}{\partial \mathbf{v}} = S = \delta(t)n_0(\mathbf{r})\gamma_0(\mathbf{v}) \tag{2}$$

Here, δ(t) is the delta function of time, **v** is the velocity, Z is the charge number, m is the mass of the ensemble particles, e is the elementary charge, and c is the speed of light. The function $\gamma_0(\mathbf{v})$ defines the initial particle velocity distribution, and the function $n_0(\mathbf{r})$ is the initial distribution in the coordinate space. By integrating equation (2), we obtain:

$$f = \Theta(t)n_0(\mathbf{r}_0)\gamma_0(\mathbf{v}_0), \qquad \Theta(t) = \begin{cases} 0, & t < 0 \\ 1, & t \geq 0 \end{cases}, \tag{3}$$

where, $\Theta(t)$ is the Heaviside function, $\mathbf{r}_0 = \mathbf{r}(t=0)$, $\mathbf{v}_0 = \mathbf{v}(t=0)$ are the initial position and velocity of particle, respectively. Having numerically calculated the trajectories of charged particles $\mathbf{r} = \mathbf{r}(\mathbf{r}_0, \mathbf{v}_0, t)$, $\mathbf{v} = \mathbf{v}(\mathbf{r}_0, \mathbf{v}_0, t)$, we can reconstruct the dependences $\mathbf{r}_0 = \mathbf{r}_0(\mathbf{r}, \mathbf{v}, t)$ and $\mathbf{v}_0 = \mathbf{v}_0(\mathbf{r}, \mathbf{v}, t)$, and, thus, find the Eulerian representation of the function *f*: $f(\mathbf{r}, \mathbf{v}, t) = \Theta(t) n_0(\mathbf{r}_0(\mathbf{r}, \mathbf{v}, t)) \gamma_0(\mathbf{v}_0(\mathbf{r}, \mathbf{v}, t))$.

The distribution function $f(\mathbf{r}, \mathbf{v}, t)$ is clearly non-stationary. To obtain its stationary value (the analogue of the distribution function calculated by the perturbation theory in the case of rare collisions [8]), we average expression (3) over time: $\langle f(\mathbf{r}, \mathbf{v}, t) \rangle = \int_0^T n_0(\mathbf{r}_0(\mathbf{r}, \mathbf{v}, t)) \gamma_0(\mathbf{v}_0(\mathbf{r}, \mathbf{v}, t)) dt / T$. At times that exceed the characteristic times of charged particles motion along the trajectory, the distribution function $f(\mathbf{r}, \mathbf{v}) = \lim_{T \to \infty} \langle f(\mathbf{r}, \mathbf{v}, t) \rangle$ no longer depends on time (convergence is provided by the conditional periodicity of trajectories) and is determined by the integrals of motion – see [7].

Further, in calculations, we use the isotropic Maxwellian particle velocity distribution $\gamma_0(\mathbf{v}) = \gamma_0(v^2) = (\pi v_T^2)^{-3/2} \exp-(v/v_T)^2$ as the source, (v_T is the thermal velocity of particles). The initial concentration of particles is considered as a function of the magnetic surface label in the unperturbed magnetic configuration $n_0(\mathbf{r}) = n_0(\Psi_0)$, namely: $n_0 = n_A\left(1-(4\cdot\Psi_0)^2\right)^2$ (n_A is the concentration of particles on the magnetic axis, the factor before Ψ_0 ensures that the plasma concentration at the boundary is zero). This choice of the source allows us to investigate changes in the distribution function of charged particles that are associated with the formation of a magnetic island. Below, calculations are given for protons with an energy of 1 keV, magnetic field on the magnetic axis $B_a = 2$ T, and aspect ratio equal three. The direction of toroidal magnetic field coincides with the direction of toroidal current.

3. Distribution function in the center of magnetic island

Let us investigate the time-averaged distribution function that is formed in the center of magnetic island in the plane $\varphi = 0$. Figure 3 shows the dependence of f on the toroidal velocity (cosine of the pitch angle) for a fixed velocity module and phase of Larmor rotation. For comparison, the figure also shows the form of distribution function in the system of nested magnetic surfaces (curve 1), which are obtained in [7]. The dashed line shows the distribution of particles from a homogeneous source with $n_0 = n_A$. In the absence of the initial concentration gradient, the distribution function (3) is a function of integral of motion (energy), and consequently, is stationary: as in the initial moment of time, it remains isotropic.

The distribution function in the center of magnetic island consists of two regions that correspond to the passing and trapped particles: the practically horizontal sections of the curve that are close to the initial distribution correspond to passing particles, and the central strongly anisotropic sloping section corresponds to trapped particles. The established form of distribution function is determined by the initial source inhomogeneity and by the trajectories of ensemble particles, see [7].

Let us compare the distribution function that is formed in the center of magnetic island (curve 2) with the distribution function on rational magnetic surface with $q = 3/2$ in the unperturbed magnetic configuration (curve 1). Their general view is very similar. The appearance of magnetic island does not change the boundaries between the trapped and passing particles, as evidenced by the same position of "spikes" on curves 1 and 2. The region of trapped particle destruction of magnetic configuration leads to the appearance of asymmetry in particle contribution with positive and negative pitch angles: in the center of magnetic island, particles with a positive toroidal velocity predominate, while in the system of nested magnetic surfaces, the fractions of particles with $v_\varphi > 0$ and with $v_\varphi < 0$ are the same (curve 1 is practically symmetric with respect to the initial value of distribution function). The main effect is related to passing particles. If in the unperturbed system the passing particles, due to the finite deviation of their trajectories from the magnetic surfaces, made an appreciable contribution to the anisotropization of the distribution function, in the center of magnetic island their contribution is negligibly small: in the region of passing particles, the distribution function is practically indistinguishable from the initial isotropic distribution of the source function.

To explain this effect, we calculate the instantaneous radial deviation of the passing particle trajectory, which originates in the center of magnetic island, from the unperturbed magnetic surface $q = 3/2$, and compare the results with particle motion in unperturbed magnetic configuration. An example of the obtained dependences for the passing particle with $v_\varphi / v = -0.9$ is shown in figure 4. One oscillation period of curve 1 corresponds to one bounce period of the passing particle. One can easily see that in the presence of magnetic island the passing particle deviates both outside and inside from the starting magnetic surface. As a result, the average radial deviation of passing particle is completely neutralized, that explains the obtained form of the distribution function.

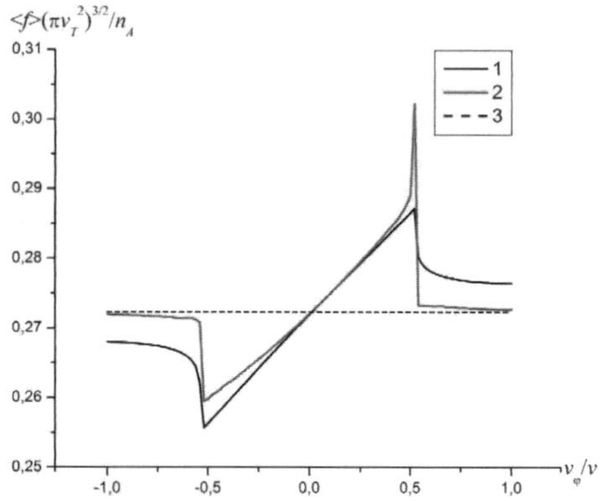

Figure 3. Normalized distribution function: (1) in the system of nested magnetic surfaces on the surface $q = 3/2$, (2) in the center of magnetic island, $\varphi = 0$. The dashed line shows the distribution from a homogeneous source with $n_0 = n_A$. The initial phase of the Larmor rotation is zero, $v = v_T$.

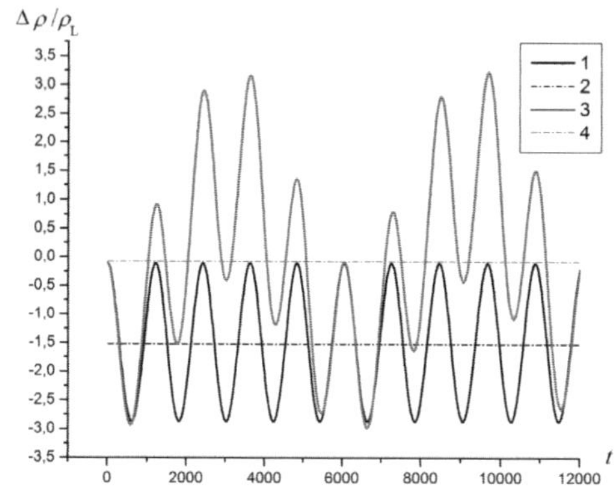

Figure 4. Time dependence of the radial deviation of the passing particle with $v_\varphi / v = -0.9$ from the unperturbed magnetic surface in the unperturbed (1) and island (2) magnetic configurations. The dashed lines show the mean values. Time is measured in cyclotron periods; the deviation is normalized on the Larmor radius.

4. Conclusions

The evolution of isotropic velocity distribution of charged particles during the collisionless motion in a toroidal-helical tokamak field with a magnetic island is considered. It is shown that the inhomogeneity of the initial particle distribution leads to anisotropization of its distribution function. In the center of magnetic island, the anisotropization is due to the motion of trapped particles. The absence of contribution from passing particles means a decrease in the efficiency of spontaneous generation of the non-induction current in magnetic island in comparison with the bootstrap effect in the system of nested magnetic surfaces [4]. The asymmetry of established distribution with a predominant contribution from trapped particles with positive pitch angles indicates an increase in the particle concentration in the center of magnetic island as a result of the considered evolution.

The publication was partly supported by the Ministry of Education and Science of the Russian Federation (agreement numbers 02.A03.21.0008 and 3.2223.2017).

References

[1] Galeev A A, Sagdeev R Z 1971 *JETP Letters* **13** 113
[2] Hinton F L, Hazeltine R D 1976 *Rev. Mod. Phys* **48** 239
[3] Bickerton R J, Connor J W, Taylor J B 1971 *Nature Phys. Sci.* **229** 110
[4] Ilgisonis V I, Sorokina E A, Yurchenko E I 2010 *Plasma Phys. Rep.* **36** (1) 1-14
[5] Sorokina E A, *Plasma Phys. Rep.* 2011 **37** (1) 19-27

[6] Ilgisonis V I, Skovoroda A A 2010 *JETP Letters* **110** (5) 890-900
[7] Ilgisonis V I, Sorokina E A 2012 *JETP Letters* **94** (9) 684-688
[8] Ben Daniel D J, Allis W P 1962 *Plasma Phys.* **4** 31-52, 79-88

IOP Conf. Series: Journal of Physics: Conf. Series **907** (2017) 012009 doi:10.1088/1742-6596/907/1/012009

Ion heat transport in ohmic plasmas of the T-10 tokamak

L A Klyuchnikov[1], V A Krupin, M R Nurgaliev, A R Nemets, I A Zemtsov, D V Ryjakov and D V Sarychev

National Research Centre "Kurchatov Institute", 1, Akademika Kurchatova pl., Moscow, 123182, Russia

[1] E-mail: lklyuchnikov@list.ru

Abstract. Results of the investigation of ion heat transport processes in the T-10 tokamak are presented. Experiments are carried out in a wide range of ohmic plasma parameters (n_e, I_{pl}, and Z_{eff}). Variation of the effective ionic charge from 3.5–4 to 1 is performed by inserting a lithium limiter. Measurements of the ion and electron temperature and density profiles are presented. The power balance approach is used for the determination of the ion heat conductivity. Modeling is performed using the ASTRA transport code. It is shown that in ohmic plasmas in the region of r/a ~ 0.5, ion heat conductivity is close to the neoclassical values, which are calculated using the NCLASS code.

1. Introduction

The necessity of achieving high efficiency plasma heating in future tokamak reactors makes it important to investigate heat transport processes. To solve this issue, energy confinement processes should be well understood. Therefore, the study of heat transport coefficients in plasmas with ohmic and auxiliary heating is performed.

On the T-10 tokamak (major radius R = 1.5 m, minor radius a = 0.3 m, toroidal filed $B_t \leq 3$ T), the study of heat transport can be performed in a wide range of ohmic parameters and with auxiliary electron-cyclotron resonance heating (ECRH) with a power up to 3 MW. Herewith, the control of light impurity (C and O) inflow using a movable lithium limiter allows us to vary the plasma effective ionic charge Z_{eff} in the range of 1 to 4.

To carry out the abovementioned investigations of heat transport, reliable measurements of ion and electron temperature and density profiles are required. On T-10, the electron temperature T_e profiles are obtained via the SXR spectra [1] and ECE diagnostics [2]. The SXR spectra measurements provide an absolute value of T_e in a particular spatial point in one tokamak discharge, while ECE diagnostics give the full profile. Electron density profiles n_e are measured using a 16-channel interferometer. Ion densities n_i and temperature T_i are measured via charge exchange recombination spectroscopy (CXRS) [3, 4].

All mentioned profiles are necessary to perform the power balance approach, which is described below. The ASTRA transport code [5] is used to determine ion heat conductivity, and the NCLASS code [6] is used for calculating neoclassical heat conductivity.

2. Power balance

The ion power balance equation in a steady state phase of discharge with ohmic heating (OH) can be written as following:

$$\mathrm{div}\left(-\chi_i^{eff}\frac{\partial T_i}{\partial r}\sum_i n_i\right)=P_{ei}-P_{CX}-P_{conv},\tag{1}$$

where, χ_i^{eff} is the effective ion heat conductivity; $P_{ei}=\sum_i \frac{T_e-T_i}{\tau_{ei}}n_e\frac{3m_e}{m_i}$ is the heat transfer from electrons to all ions, τ_{ei} is the electron-ion collision time; $P_{CX}=\frac{3}{2}(T_i-T_0)n_d n_0\langle\sigma v\rangle_{CX}$ are the heat losses due to charge exchange of deuterons (n_d) on deuterium atoms (n_0) with temperature T_0, $\langle\sigma v\rangle_{CX}$ is the charge exchange rate coefficient; $P_{conv}=\frac{5}{2r}\frac{\partial}{\partial r}(r\cdot T_i\cdot\Gamma_i)$ are the heat losses due to ion transport, and Γ_i is the ion flux. Thus, the effective coefficient of ion heat conductivity can be determined from (1):

$$\chi_i^{eff}=-\frac{1}{r}\frac{\int_0^r(P_{ei}-P_{CX}-P_{conv})rdr}{\nabla T_i\cdot\sum_i n_i}\tag{2}$$

Charge exchange and convective losses can be significant at the plasma periphery, while in the central plasma region ($\rho < 0.8$), they can be ignored. To determine χ_i^{eff}, the radial profiles of T_e, T_i, n_e, and n_i need to be carefully measured.

3. Measurements of electron and ion temperatures

Ion temperature profiles on T-10 are measured via the CXRS diagnostics, which is described in [3, 4]. Measurements are performed via Doppler broadening of the C^{5+} 5291 Å CXRS-line.

Examples of measured profiles in discharges with a different plasma current are shown in fig. 1. The CXRS diagnostics allow to investigate the T_i values and shape changes in a wide range of plasma parameters.

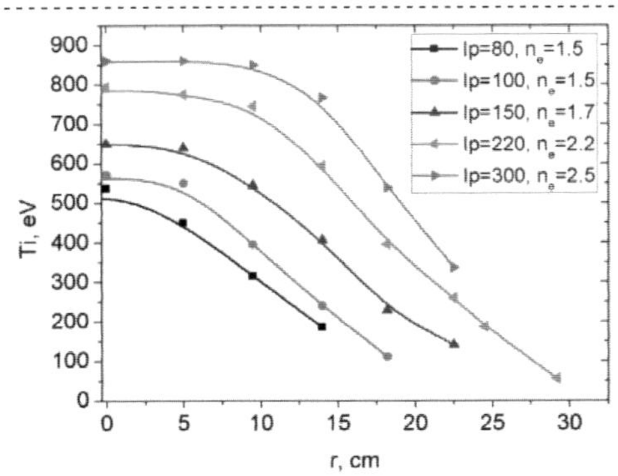

Figure 1. Ion temperature profiles measured via CXRS on T-10 in discharges with ohmic heating

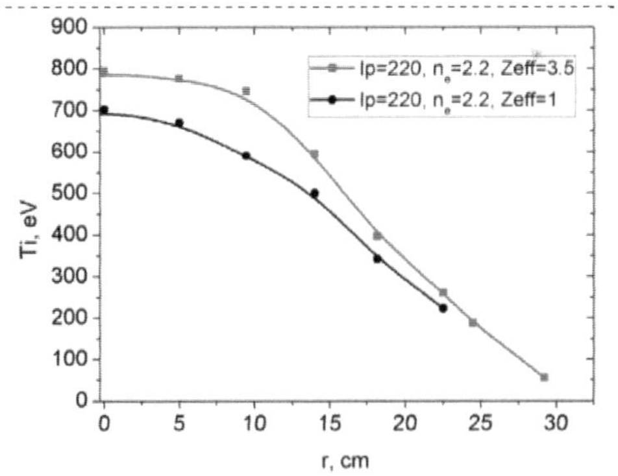

Figure 2. Ion temperature profiles measured in plasmas with different Z_{eff} values

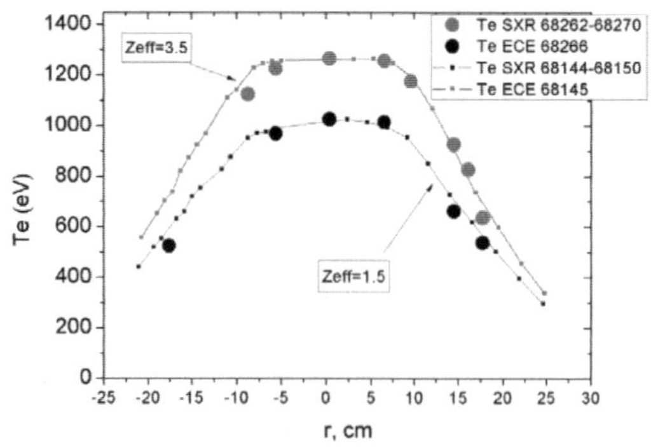

Figure 3. Electron temperature profiles measured via the SXR spectra and ECE diagnostics in plasmas with different Z_{eff} values

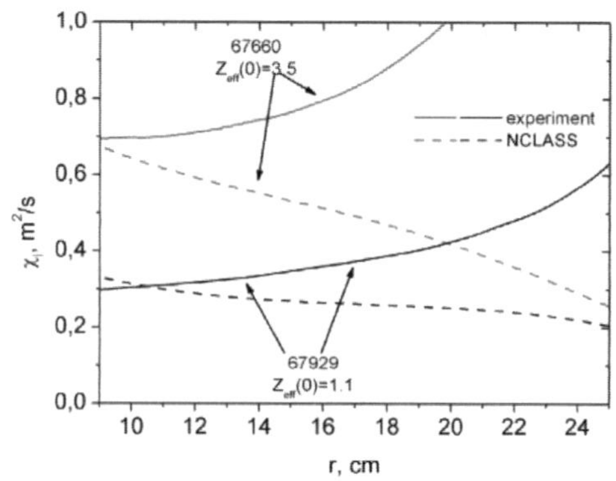

Figure 4. Ion heat conductivity determined from the power balance (solid lines), and neoclassical values (dashed lines) that are calculated via NCLASS in discharges with low and high Z_{eff} values

The ion temperature profiles, which are shown in fig. 2, are measured in discharges with high $Z_{eff} \approx 3.5$ and low $Z_{eff} \approx 1$, which are achieved after vacuum vessel lithization. In high Z_{eff} OH plasmas contaminated by light impurities, the obtained values of T_i are higher than in the low Z_{eff} plasma case. However, the shape of $T_i(r)$ changes insignificantly.

To determine heat transport coefficients, reliable measurements of $T_i(r)$ and accurate measurements of $T_e(r)$ are required because ions are heated via electron-ion collisions. When comparing $T_e(r)$ that is measured in discharges with high and low Z_{eff} using SXR spectra and ECE diagnostics, one can observe that the electron temperature profiles, which are measured using both diagnostics, match well. The values of T_e in the central region are significantly higher in discharges that are contaminated by light impurities than in plasma with $Z_{eff} \approx 1.5$. It is necessary to note that the experimental $T_e(r)$ profiles together with the measured Z_{eff} can be used to describe the loop voltage and saw-tooth inverse radius in the ASTRA code.

4. Experimental results

Experiments with wall lithization are performed on the T-10 tokamak to vary the content of light impurities. In the discharge with a plasma current $I_{pl} = 220$ kA and line-averaged electron density $\bar{n}_e = 2.5 \cdot 10^{19}$ m^{-3}, the values of Z_{eff} are changed in the range of 3.5–4 to 1.

Ion heat conductivity coefficients, which are determined via the power balance approach in discharges with high and low Z_{eff}, are presented in fig. 4. It is shown that with an increase of the Z_{eff} value, the increase of the experimental and neoclassical transport coefficients is observed. Herewith, the values of experimental heat conductivity are close to the neoclassical value in the central area $\rho \leq 0.5$. However, at the plasma periphery, χ_i^{eff} increases drastically, while the neoclassical value decreases. This can be an evidence of the existence of anomalous part of the ion heat flux, which is dominant at the plasma edge.

As shown above, the increase of Z_{eff} from 1 to ~4 leads to the insignificant increase of T_i to 10–15% together with a strong decrease of the total ion density $\sum_i n_i$ up to 50% due to dilution of the main gas by light impurities (C and O). Thus, the total energy of plasma ions decreases with an increase of the Z_{eff} value. In turn, the heat transfer from electron to all ions changes slightly despite the growth of T_e because of the $\sum_i n_i$ decrease.

Wall lithization allows to achieve the values of Z_{eff} ~ 1–1.3 as well as high values of line-averaged electron density (up to 5.6 · 10^{19} m^{-3}) in ohmic discharges with I_{pl} = 220 kA. In fig. 5, the dependence of effective ion heat conductivity at ρ = 0.5 on electron density is shown. The presented experimental transport coefficients are in good agreement with neoclassical ones, which are calculated using the NCLASS code in the entire range of $\bar{n}_e$.

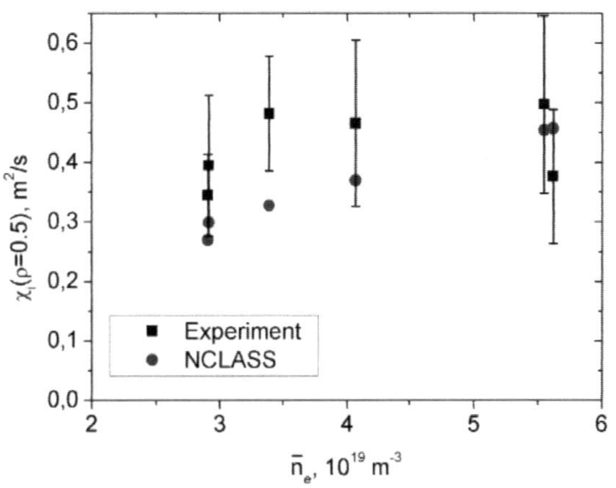

Figure 5. The dependence of $\chi_i^{eff}(\rho = 0.5)$ on plasma density in discharges with I_{pl} = 220 kA

5. Conclusion

The electron, ion temperature and density diagnostics of the T-10 tokamak provides data that is necessary for the ion heat transport analysis. Measurements are performed in discharges with various values of electron density and effective ionic charge.

The ion heat conductivity values, which are determined via the power balance approach, are similar to neoclassical values in the range of ρ ≤ 0.5. The experimental and neoclassical heat conductivities increase with an increase in Z_{eff}. However, the experimental profile of $\chi_i^{eff}(r)$ increases at the plasma periphery where it exceeds the neoclassical coefficient. Therefore, it can be concluded that anomalous heat transport dominates neoclassical transport at the plasma edge.

Analysis of the electron heat transport will be performed in future works. Moreover, regimes with auxiliary ECR heating will be also investigated. It is important to consider the data of plasma density fluctuations [7] and plasma potential fluctuations [8] and compare them with our results to study the relations of turbulent processes and plasma transport.

Acknowledgments

The authors are grateful for the financial support to the Russian Foundation for Basic Research (Project № 16-32-00148 mol_a). Modernization of ion temperature and density diagnostics is performed with the funding support from the Russian Science Foundation (Project No. 14-22-00193). Experiments with a lithium limiter are supported by The State Atomic Energy Corporation ROSATOM (contract № N.4x.241.9B.17.1011, 20.02.2017)

References

[1]. Trukhin V M et al 2013 *Prob. of atomic sci. and tech., ser. Therm. Fusion* **36** 1 78-85
[2]. Poznyak V I et al 2012 *EPJ Web of Conferences* **32** 03010
[3]. Krupin V A et al 2014 *Prob. of atomic sci. and tech., ser. Therm. Fusion* **37** 4 60-70
[4]. Klyuchnikov L A et al 2016 *Review of Sci. Instr.* **87** 053506.
[5]. Pereverzev G V and Yushmanov P N 2002 *Max-Plank IPP report* 5/98
[6]. Houlberg W et al 1997 *Phys. Plasmas* **4** 3230
[7]. Vershkov V A et al 2005 *Nucl. Fusion* **45** 10 203-26
[8]. Melnikov AV et al 2006 *Plasma Phys.Control. Fusion* **48** S87

IOP Conf. Series: Journal of Physics: Conf. Series **907** (2017) 012010 doi:10.1088/1742-6596/907/1/012010

Study of thermal quench of discharge using a massive gas injection into the plasma of T-10 tokamak

V G Kapralov[1,3], M M Dremin[2], H A Harfush[1], S V Krylov[2], V G Skokov[1], A E Borovov[1], V V Elagin[1], K S Sedov[1] and I A Sharov[1]

[1] SPbPU, Saint-Petersburg, RF
[2] NRC «Kurchatov Institute», Moscow, RF

[3] E-mail: kapralov@phtf.stu.neva.ru

Abstract. The report presents the results of simulations of plasma disruption in the T-10 tokamak using the ASTRA code, including discharges with a disruption being initiated by the massive gas injection (MGI). This study focuses on the development phase of thermal quench, and the effect of different channels of heat loss during quench evolution is examined. It is shown that the initial phase of the slow phase of thermal quench can be described by the selection of sources and transport coefficients.

1. Introduction

The design and putting into operation of increasingly powerful tokamaks creates a problem of preventing major plasma disruptions and controlled quench of discharge. It should be noted that the physics of the processes that occur during a disruption depends more on the geometry of tokamak magnetic system and the plasma confinement regimes implemented there than on the amount of stored energy in the plasma discharge. This makes it possible to perform preliminary studies on this subject using small and medium tokamaks with less risk of damage, because they have a lower plasma energy content.

To study the various scenarios of initiation and evolution of plasma disruption and suppression of the runaway electron beams, it is necessary to use a variety of active systems that affect plasma in order to to test various scenarios of controlled quench of the plasma discharge. At the T-10 tokamak, it is possible to use the following active systems: firstly, it is the main control system for executing control programs for the plasma current, magnetic fields and operation of piezoelectric valves for the working gas injection, and secondly, stationary and movable (positionable) pulsed gas valves, impurity pellet injector (the latter is now equipped with a chord pellet injection system), as well as an electron cyclotron resonance plasma heating (ECRH) system, movable and lithium limiters [1]. It should be noted that the unique systems of movable MGI valve [2] and chord pellet injection allow to study disruption parameters as a function of gas source position relative to the plasma edge and the impact parameter of pellet injection, respectively, and compare the scenarios of co- and counter-pellet-injection ("co-" means injection along the direction of plasma rotation, "counter-" corresponds to injection in the opposite direction against the direction of plasma rotation).

On the Tore Supra tokamak with a limiter, the injection of approximately 0.1 moles of helium was carried out for 4 ms in the experiments on quenching of ohmic plasma discharge with a current of up to 1.2 MA [3]. This amount of helium was calculated numerically based on experimental results of the

IOP Conf. Series: Journal of Physics: Conf. Series **907** (2017) 012010 doi:10.1088/1742-6596/907/1/012010

FTU tokamak [4] using an assumption that impurity injection into steady-state plasma should suppress and during thermal quench reduce the generation of runaway electron beams.

The decrease in runaway electron current to the wall was observed using the response of HXR detectors. The HXR emission increases simultaneously with the appearance of photo-neutrons, which are produced when runaway electrons reach the first wall of the facility. It is observed that in discharges with helium injection, the number of runaway electrons, which reached the plasma-facing elements of the facility, is less (by 2-3 orders of magnitude) than in experiments with plasma disruptions without MGI [3].

In addition to massive helium injection, experiments with an ordinary argon injection were performed at the Tore Supra tokamak. With the massive helium injection, smaller eddy currents formed, and the plasma current decreased slower than in the case of argon injection, which led to lower loads on the constructional elements of the facility during disruptions. Based on the results of experiments on the Tore Supra tokamak, it has been demonstrated that the massive helium injection does not produce plasma pollution, and cleaning of the first wall of the device became easier [3].

2. Experiment

For the numerical modeling of plasma thermal quench, which is caused by the impurity injection, a discharge #66640 with a thermal and current quench in the T-10 tokamak with a limiter was chosen.

In this experiment, the injection of argon at a pressure of 3 bar was carried out using a gas valve (Figure 1), which is located 30 mm from the last closed magnetic surface, at the time of 834 ms from the beginning of discharge.

Prior to the injection of argon impurity, a steady-state ohmic operating mode of the tokamak was achieved with an average electron concentration of 2.5×10^{19} m^{-3}, a current in the plasma of $I = 250$ kA, and an average loop voltage of $U = 1.0$ V. When argon was injected, the plasma periphery cooled, the current was displaced to the central part, and the electrical conductivity of plasma decreased. The ionization of impurity led to an increase in the average electron density and subsequent plasma discharge disruption because of exceeding the plasma density limit.

The electron density profile before the injection, the characteristics of the T-10 tokamak and experimentally measured plasma parameters were taken as the initial data for modeling. The initial electron temperature profile was chosen to be parabolic and normalized to correspond with the value of loop voltage $U = 1.0$ V.

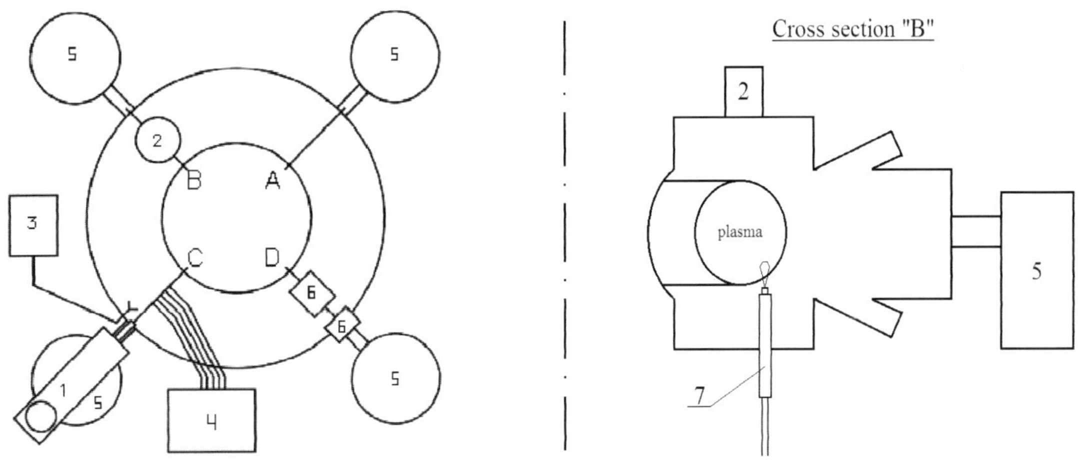

Figure 1. T-10 tokamak scheme (top view and cross section B): 1 – hydrogen pellet injector, 2 – impurity pellet injector, 3 – electron temperature diagnostics, 4 – microwave interferometer, 5 – gyrotrons, 6 – SXR diagnostics, 7 – movable gas valve for MGI.

In Figures 2 and 3, solid lines represent the experimental responses of electron temperature to the chord passing through the radius of plasma discharge r = 0.21 m and loop voltage, respectively. They represent the early slow phase of the thermal quench. In the T-10 tokamak, this phase lasts for 100-200 μs. The outer regions of plasma cool down, and the effective charge of plasma and radiation losses increase during this time. In the non-central regions of plasma, the electrical conductivity decreases, which leads to the increase in loop voltage and the displacement of current into the central part of the plasma discharge. In the simulation, it was expected to obtain a qualitatively similar evolution of the corresponding values after the impurity injection.

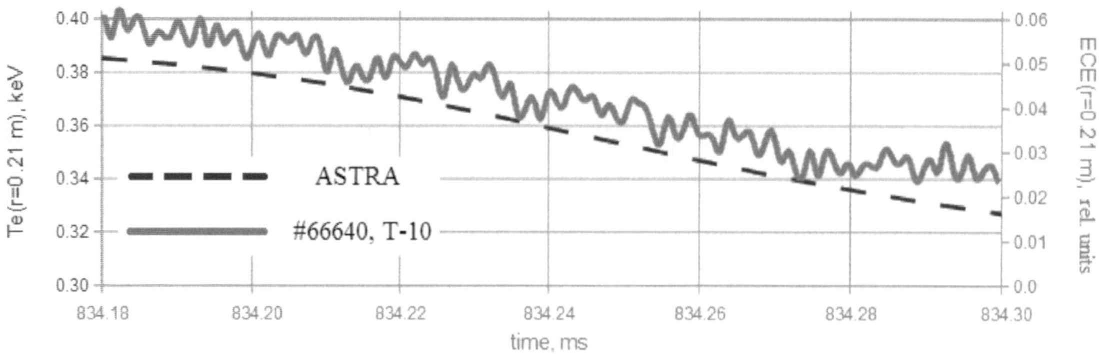

Figure 2. Electron temperature at the minor radius r=0.21 m – modeling and experiment (injection – 834 ms).

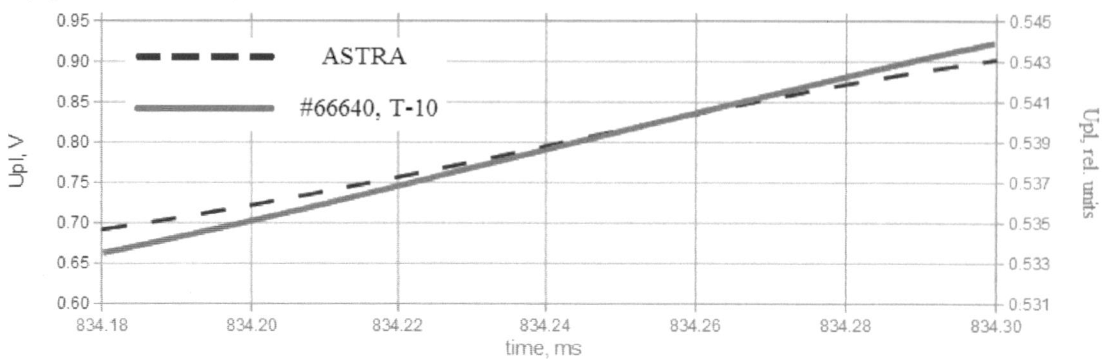

Figure 3. Loop voltage – modeling and experiment (injection – 834 ms).

3. Modeling

Numerical study on the influence of injected impurities on the plasma parameters of the tokamak during the first stage of thermal quench was carried out using the ASTRA code [5], which numerically solves the system of transport equations (1) for the electron density n_e, electron temperature T_e, ion temperature T_i and poloidal flux θ.

The first equation describes the balance of electrons (the number of particles conservation law), the second and third equations describe the energy balance (the energy conservation law) for electrons and ions, and the fourth equation describes Ohm's law for a plasma discharge. The system (1) is complemented by the appropriate initial and boundary conditions of the first kind [5]. In the system (1), ρ is the effective minor radius; t is the time; V is the volume of plasma; $V' = \partial V/\partial \rho$, B_0 and $\dot{B}_0$ are the magnetic field and its time derivative at the center of the plasma discharge, respectively; Γ_e and Γ_i are the electron and ion fluxes, respectively; q_e and q_i are the corresponding heat fluxes; n_e and n_i are the electron and ion concentrations, respectively; S_e, P_e and P_i are the terms that take into account electron sources, electron and ion heating sources, respectively; $\sigma_{||}$ is the conductivity of plasma in the direction parallel to the magnetic field; $J = I/(R_0 \cdot B_0)$, where R_0 is the radius of the tokamak vacuum chamber, I is the plasma current; j_{BS} and j_{CD} are the averaged bootstrap current densities due to the electron cyclotron resonance heating of plasma, and the current caused by external sources, respectively; $G_2 = \langle(\nabla\rho/r)^2\rangle \cdot V'/(4\pi^2)$.

$$\begin{aligned}
&\frac{1}{V'}\left(\frac{\partial}{\partial t}-\frac{\dot{B}_0}{2B_0}\frac{\partial}{\partial\rho}\rho\right)(V'n_e)+\frac{1}{V'}\frac{\partial}{\partial\rho}\Gamma_e=S_e,\\
&\frac{3}{2}(V')^{-5/3}\left(\frac{\partial}{\partial t}-\frac{\dot{B}_0}{2B_0}\frac{\partial}{\partial\rho}\rho\right)\left[(V')^{-5/3}n_eT_e\right]+\frac{1}{V'}\frac{\partial}{\partial\rho}\left(q_e+\frac{5}{2}T_e\Gamma_e\right)=P_e,\\
&\frac{3}{2}(V')^{-5/3}\left(\frac{\partial}{\partial t}-\frac{\dot{B}_0}{2B_0}\frac{\partial}{\partial\rho}\rho\right)\left[(V')^{-5/3}n_iT_i\right]+\frac{1}{V'}\frac{\partial}{\partial\rho}\left(q_i+\frac{5}{2}T_i\Gamma_i\right)=P_i,\\
&\sigma_{||}\left(\frac{\partial\psi}{\partial\rho}-\frac{\rho\dot{B}_0}{2B_0}\frac{\partial\psi}{\partial\rho}\right)=\frac{JR_0}{\mu_0\rho}\frac{\partial}{\partial\rho}\left(\frac{G_2}{J}\frac{\partial\psi}{\partial\rho}\right)-\frac{V'}{2\pi\rho}\left(j_{BS}+j_{CD}\right)
\end{aligned}\tag{1}$$

The injection of gas (argon) was included in the system (1) via S_{imp}, P_{rad} and P_{Br} in the sources S_e and P_e:

$$S_e = S_{ion} - S_{rec} + S_{imp}, \tag{2}$$

$$P_e = P_{OH} - P_{ei} - P_e^N - P_{rad} - P_{Br}, \tag{3}$$

where, P_{OH} is the ohmic heating power; P_{ei} and P_e^N describe the energy exchange due to the interaction of electrons with ions and neutral particles, respectively; P_{rad} is the power radiated due to impurity injection; P_{Br} is bremsstrahlung; S_{ion} and S_{rec} describe the processes of ionization and recombination, respectively; S_{imp} is the ionization/recombination of the injected impurity atoms.

The formulas for S_{ion}, S_{rec}, P_{OH}, P_{ei} and P_e^N were originally built in the ASTRA code [5]. However, for calculating S_{imp}, P_{rad} and P_{Br}, the appropriate subroutines were developed according to the following expressions:

1) $S_{imp} = n_{imp}\cdot\partial\langle Z_{imp}\rangle/\partial t = n_{imp}\cdot\partial\langle Z_{imp}\rangle/\partial T_e\cdot\partial T_e/\partial t$, where the derivative $\partial\langle Z_{imp}\rangle/\partial T_e$ was calculated using the average-ion model [6, 7];
2) $P_{rad} = n_e\cdot n_{imp}\cdot L_z$, where $[P_{rad}] = erg\cdot cm^{-3}\cdot s^{-1}$, $[n_e] = [n_{imp}] = cm^{-3}$ [7];
3) $P_{Br} = 5.35\cdot 10^{-37}\cdot Z^2_{eff}\cdot n_e\cdot n_{imp}\cdot T_e^{1/2}$, where $[P_{Br}] = W\cdot m^{-3}$, $[T_e] = keV$ [8].

According to [5], the ASTRA code can include up to three auxiliary transport equations:

$$\frac{\partial}{\partial t}\left(V' f_j\right)=\frac{\partial}{\partial\rho}\left[V'\langle(\nabla\rho)^2\rangle\left(D_j\frac{\partial f_j}{\partial\rho}-u_jf_j\right)\right]+V' S_j,\quad j=1,2,3 \tag{4}$$

with the appropriate initial and boundary conditions. The first of these equations was used to describe the influence of impurity density n_{imp} on transport processes, i.e., $f_1 = n_{imp}$. Then, index 1 corresponds to the quantities that describe the impurity behavior. This allowed us to determine for the impurity its sources S_1, diffusion coefficient D_1 and pinch velocity u_1, and to simulate the evolution of concentration of argon particles after injection. Similar to [6], the values of D_1 and u_1 were set as for the electrons, and source S_1 was set to 0. The condition of the third type (conservation of the number of impurity particles in the plasma volume) was chosen as the boundary condition for the impurity:

$$\left.\left(-D_1\frac{\partial f_1}{\partial\rho}+u_1f_1\right)\right|_{\rho=\rho_s}=0 \tag{5}$$

As a result of the simulation, plots of the electron temperature at the radius of plasma discharge r = 0.21 m and loop voltage were obtained, which qualitatively coincide with the plots of the real discharge #66640 in the T-10 tokamak with a limiter (Figures 2 and 3). Figure 4 demonstrates the profiles of electron temperature, which were obtained as a result of modeling the disruption using the ASTRA code. From Figure 4, it is clear that after the compression of plasma into the central regions, plasma mixing does not occur, therefore. This physical phenomenon, which is an essential part of the second stage of a disruption, cannot be described using the model that was discussed in this article.

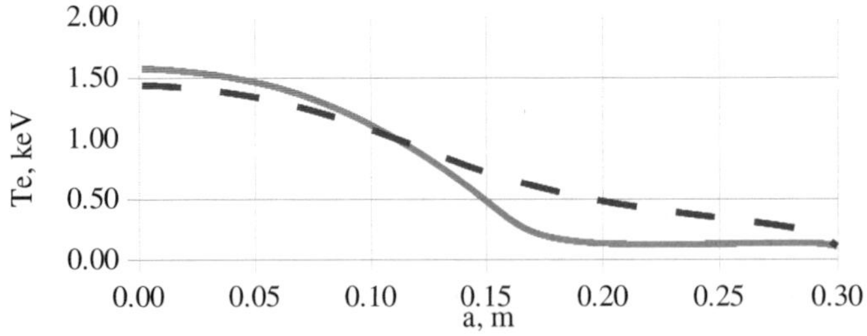

Figure 4. Electron temperature T_e profiles before (dashed line) and after (solid line) the slow phase of the thermal quench.

4. Conclusions

The report presents the results of a numerical simulation of a discharge disruption in the T-10 tokamak using the ASTRA code, including discharges with the initiation of a discharge disruption using a massive pulsed gas injection. The main attention is paid to the phase of the evolution of thermal plasma quench and the effect of various heat loss channels during it. It is shown that the initial slow phase of the development of thermal plasma quench can be described by selecting sources of impurity and transport coefficients. The comparison of the results of modeling with the experiment demonstrates that the proposed approach of taking into account the injected gas in the ASTRA code by solving the auxiliary transport equation allows to qualitatively simulate the evolution of plasma parameters during the first stage of a thermal quench, which is caused by the impurity injection.

Acknowledgments

The authors are thankful to the staff of the T-10 tokamak for the provided data and support. The work was carried out with the partial support from the State Corporation ROSATOM, contract No.H.4x.241.9B.17.1011.

References

[1] Dremin M M et al 2012 *Problems of Atomic Science and Technology, Ser. Thermonuclear Fusion* **4** 58

[2] Kapralov V G et al 2016 *Proc. XLIII Zvenigorod Int. Conf on Plasma Physics and Controlled Fusion* (Zvenigorod, 2016), (Moscow: PLAZMAIOFAN) 123

[3] Martin G, Sourd F, Saint-Laurent F, Bukalossi J and Eriksson L G 2004 Disruption Mitigation on Tore Supra *20th IAEA Fusion Energy Conf. 2004* (Vilamoura) EX/10-6Rc 36

[4] Esposito B et al 2003 *Physics of Plasma* **10** 2350

[5] Pereversev G V and Yushmanov P N 2002 ASTRA - Automated System for TRansport Analysis Preprint IPP 5/98

[6] Timokhin V M 2002 *Study of Plasma Quench and Fast Electrons in Magnetic Plasma Confinement using Pellet Injection* (St. Petersburg: St. Petersburg State Polytechnic University)

[7] Post D E, Jensen R V, Tarter C B, Grasberger W H and Lokke W A 1977 Steady-state radiative cooling rates for low-density, high-temperature plasmas *Atomic Data and Nuclear Data Tables* 20 PPPL-1352 pp 397-439

[8] Wesson J 2004 *Tokamaks* (Oxford: Clarendon Press)

Analysis of discharge scenarios in the T-15 tokamak

D Yu Sychugov[1], I V Zotov[1,5], N V Kasyanova[2], A V Melnikov[2,3], A V Sushkov[2], A D Sadykov[4] and G V Shapovalov[4]

[1] Faculty of Computational Mathematics and Cybernetics, Lomonosov Moscow State University, 119991 Moscow, Russia
[2] National Research Centre "Kurchatov Institute", 123182 Moscow, Russia
[3] National Research Nuclear University MEPhi, 115409 Moscow, Russia
[4] Institute of Atomic Energy of NNC RK, 071100 Kurchatov, Kazahstan

[5] E-mail: iv-zotov@cs.msu.ru

Abstract. Currently, all of the main elements of the T-15 design have acquired their specifics, and, therefore, it is possible to conduct a sufficiently detailed analysis of future experiments. In this paper, we simulated the stage of current increase and the beginning of transition to a configuration with a divertor. We were interested in the degree of correlation between the scenario calculations, which were carried out using various numerical codes, the existence or absence of the problem with vertical plasma instability at this stage, and the accuracy of restoration of the plasma boundary, which depends on the error of magnetic sensors. As a result of calculations, it was possible to show that the stage of discharge, which was considered by us, is quite real from the point of view of its realization.

1. Introduction

The current construction of the T-15 tokamak makes urgent the task of analyzing the basic scenarios of discharge, which are supposed to be performed at this facility. The analysis of discharges at the T-15 installation at the stationary stage was carried out in [1, 2]. It was shown that such states can be supported by feedbacks. It was also shown that the magnetic diagnostic system can restore the plasma boundary with a specified accuracy, if the measurement error does not exceed 1–3%.

Meanwhile, the analysis of stationary states alone is inadequate. An example of an "unsuccessful" (unrealizable) scenario is given in [3], during which, because of an early stretching of the plasma pinch, a configuration appeared, whose vertical instability could not be suppressed by the feedback system.

Currently, all of the main elements of the T-15 design have acquired their specifics [4], and, therefore, it is possible to conduct a fairly realistic analysis of the stages of current increase and exit to the stationary stage during a discharge at the T-15 facility. The purpose of this work was to carry out this analysis. The stage of current increase and the beginning of transition to a configuration with a divertor was chosen for the analysis. For the first time, this analysis was carried out in [7], and the calculations were conducted using the widely known DINA code [8–9]. Currently, a new open-access computing resource (nfusion.cs.msu.ru), which integrated the TOKSCEN (equilibrium, vertical stability and evolution of plasma) [5] and RPB (restoration of the plasma boundary) [6] modules, was developed by a group of researchers from the Moscow State University and the IAE of the Republic of Kazakhstan. The purpose of this work was to perform a detailed analysis of the stage of current increase in the T-15 installation. We were interested in the following questions:

1) Comparison of the calculation results of the control points of the scenario, which was carried out according to the DINA and TOKSCEN codes;

2) Analysis of vertical plasma instability at the studied stage of discharge;

3) Restoration of the plasma boundary error as a function of the sensor error in the magnetic field measurements.

As a result of the calculations, it was possible to show that the stage of discharge considered by us is real from the point of view of its realization.

2. Results of the calculations

We analyzed the stage of current increase and pulling of the plasma column in vertical direction during the evolution of discharge in T-15 (discharge time from 0.7 to 1.2 sec) [7]. Several moments of time were selected within this interval, and each one was determined via the following: 1) the degree of compliance for the two different DINA and TOKSCEN codes; 2) the possibility of suppressing the vertical instability of the plasma; and 3) the accuracy of determining its boundary, taking into account the error in the signals on the sensors.

Figures 1–3(a) consistently represent the discharge moments t = 0.7, 1.0 and 1.19 s. It is important to note that the degree of correlation between the calculation results from the DINA and TOKSCEN codes is very good. There is a coincidence of up to three significant figures, both in magnitude of the currents and in terms of the geometric characteristic of the plasma pinch. Figures 1(b) and 3(b) show in different colors the accuracy of the boundary restoration and the separatrix with a relative error of signals at the sensors equal to 1%. A system of 36 two-component field sensors was used. It can be seen that with an error level of less than 1%, the error of determining the plasma boundary is not more than 2 cm, which is acceptable. However, an increase in the measurement error of the field to 3% leads to a serious deterioration in the determination of the accuracy of the boundary to a value of 3.5 cm, which is an unacceptable value [Fig.2(b)]. To maintain an acceptable accuracy of 2 cm, it is necessary to increase the number of magnetic sensors to 44. Note that the position of the "whiskers" of the separatrix and the x-point itself are controlled at a much higher accuracy – approximately 2.5 mm. This ensures the effective control of the divertor performance. As for the vertical instability, for the 0.7 and 1.0 s moments of time, it is absent, which is explained by the small ellipticity of the pinch cross section. Only at time t = 1.19 s, there is an instability with a small increment equal to 1.554 s^{-1} (τ = 643.394 ms), which can be easily suppressed by the feedback.

3. Conclusion

As a result of calculations, it was possible to show that the stage of discharge, which is considered by us, is real from the point of view of its realization. Specifically:

1) the equilibrium calculations performed using two different DINA and TOKSCEN codes, coincided in both geometric characteristics and holding currents up to the third significant digit;

2) the vertical instability of plasma at this stage of the scenario is either completely absent or its increment reaches a non-dangerous value of the order of 1.5 s^{-1}, which is easily suppressed by the feedback. This result, unlike the "unfortunate" scenario [3], is explained by the fact that vertical pulling of the plasma cord starts to occur at the time when the current in the plasma reaches a significant value of approximately half of the maximum possible value, and the plasma fills a significant part of the chamber [Fig. 1(a)].

3) the plasma boundary at a given stage of discharge is controlled with an acceptable accuracy of less than 2 cm, provided that the relative error in determining the magnetic field of the sensors corresponds to the declared nominal value (1%). However, the error in determining the boundary becomes critically large (approximately 3.5 cm), if it turns out that the error in determining the fields by the sensors is higher than approximately 3%. The performed calculations show that the stage of discharge considered by us is realistic from the point of view of its realization.

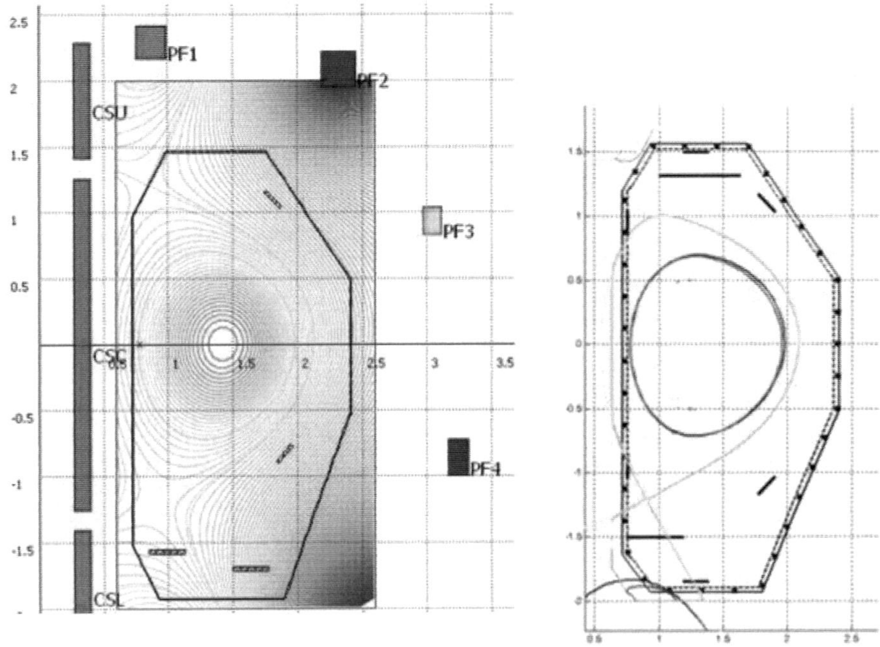

a) b)

Figure 1. (a) Stage of current rise in the T-15 installation. The moment of time t = 0.7 s, total plasma current $I_p = 0.91$ MA, large radius $R_{mid}95 = 1.49$ m, small radius a = 0.60 m, ellipticity of the plasma column cross section k = 1.17, triangularity $\Delta = 0.11$, and poloidal beta $beta_p = 0.1$. (b) Reconstruction of the plasma boundary (red and blue lines) and separatrix (green and yellow lines) with 36 two-component magnetic field sensors (black squares), the relative error of measurements is 1%.

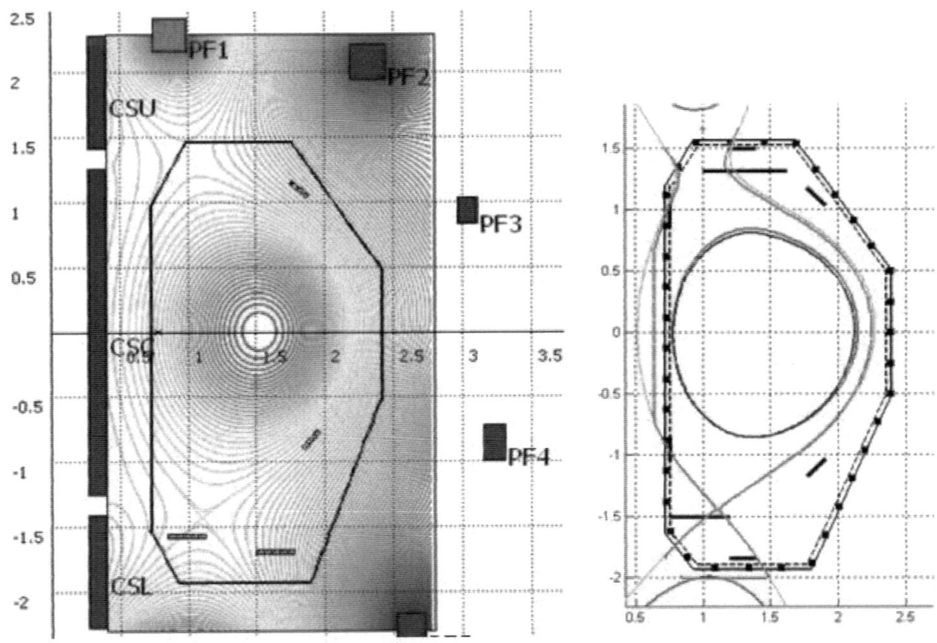

a) b)

Figure 2. (a) Magnetic configuration of T-15 at the moment of time t = 1.0 s, $I_p = 1.36$ MA, $R_{mid}95 = 1.463$ m, a = 0.68 m, k = 1.25, $\Delta = 0.1$, $beta_p = 0.14$. (b) The exact and reconstructed boundary of the plasma (blue and red lines), and the exact and reconstructed separatrix (green and purple lines) at the same time (measurement error 3%), 36 two-component magnetic field sensors (black squares).

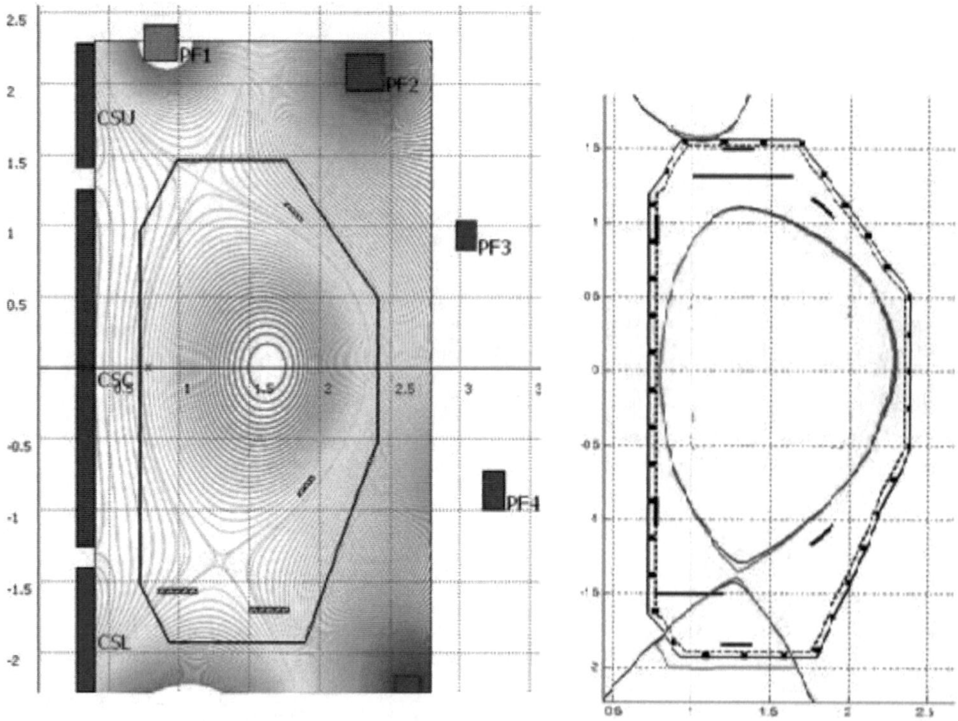

a) b)

Figure 3. (a) Magnetic configuration of T-15 at the moment of time t = 1.19 s, I_p = 1.58 MA, $R_{mid}95$ = 1.416 m, a = 0.63 m, k = 1.48, Δ = 0.11, $beta_p$ = 0.13. (b) The exact and reconstructed boundary of the plasma (separatrix – red and blue lines) at the same time, the relative level of measurement error of 1%, 36 two-component magnetic field sensors (black squares).

Acknowledgments

This work was carried out with the partial support from the State Corporation ROSATOM (Contract No. H.4x.241.9Б.17.1011), as well as RFBR grants No. 17-07-00544-a and No. 17-07-00883-a. The work of AVM was partly supported by the Competitiveness Program of NRNU MEPhi.

References

[1] Kasyanova N V, Andreev V F, Melnikov A V, Sushkov A V, Sychugov D Yu, Sadykov A D 2014 *41st EPS Conf. on Plasma Phys. and Contr. Fus. (Berlin, Germany, 2014)* P4.072
[2] Zotov I V, Melnikov A V, Sychugov D Yu, Lukash V E, Khayrutdinov R 2016 *43rd EPS Conf. on Plasma Phys. and Contr. Fus. (Leuven, Belgium, 2016)* P2.035
[3] Kakurin A M et al 2005 *Problems of Atomic Science and Technology. Ser. Thermonuclear Fusion* **28** 4 53-75
[4] Melnikov A V et al 2015 *Fusion Engineering and Design* **96-97** 306-310
[5] Sadykov A D, Sychugov D Yu, Shapovalov G V, Chektybaev B Zh, Skakov M K and Gasilov N A 2015 *Nuclear Fusion* **55** 4 55043017
[6] Belov A G, Zotov I V, Sychugov D Yu 2012 *SCET2012 - Spring World Congress on Engineering and Technology (Xi'an, China, 2012)* 278-280 (http://www.scirp.org)
[7] Dokuka V N, Kavin A A, Lukash V E, Sokolov M M, Khayrutdinov R R, Khayrutdinova V A 2014 *Problems of Atomic Science and Technology. Ser. Thermonuclear Fusion* **37** 3 56

[8] Khayrutdinov R R and Lukash V E 1993 *J. Comput. Physics* **109** 193-201
[9] Khayrutdinov R R, Lister J B, Dokuka V N, Duval B P, Favez J-Y, Lukash V E, Raju D 2003 *30th EPS Conf. on Plasma Phys. and Contr. Fus. (St.-Petersburg, Russia, 2003)* P-3.163. *Preprint* LRP 766/03 53-56
[10] Dnestrovskij A, Voss G, Sychugov D, Lukash V, Khayrutdinov R 2007 *34 EPS Conf. on Plasma Phys. and Contr. Fus. (Warsaw, Poland, 2007)* P-1.101

Preliminary results of divertor modelling for DEMO-FNS reactor

A S Kukushkin[1,2,4], V Yu Sergeev[3] and B V Kuteev[1,2]

[1] National Research Center "Kurchatov institute", Moscow, Russia
[2] National Research Nuclear University "MEPhI", Moscow, Russia
[3] Peter the Great St. Petersburg Polytechnic University, St. Petersburg, Russia

[4] E-mail: ank755@gmail.com

Abstract. The first results of divertor modelling for the DEMO-FNS divertor are presented. The divertor geometry open towards the private flux region is found important for improving the pumping conditions and reducing the power loading on the targets.

A "conventional" tokamak reactor is presently seen as a huge machine with high power and enormous cost. Realization of such a project is a real challenge [1]. The most demanding goal of this reactor is economically efficient production of energy, which implies operation with ignition or close to it, the fusion gain being Q > 10. However, this requirement can be reduced down to Q $\gtrsim$ 1 if the hybrid, fusion plus fission, scheme is considered. In this case, the tokamak plays the role of a fusion neutron source (FNS) and the efficiency of energy production is ensured by a sub-critical fission blanket [2, 3]. Besides energy production, such a scheme could also be used for breeding the fissile fuel, for transmutation of minor actinides from nuclear waste, or for tritium production [3].

A concept of a compact FNS based on the spherical tokamak (ST) with the aspect ratio A = $R/a \gtrsim 1$ was proposed in [4, 5]. High β limit and good energy confinement demonstrated on ST [6] offer a possibility of building a tokamak with reactor-relevant plasma and a major radius $R \sim 1.5$ m [7, 5]. However, the neutron shielding of the central post that closes the toroidal field coils in ST, necessary to maintain its electrical conductance during the machine lifetime, requires an increase of R to about 3 m, resulting in a machine with the fusion power $P_{\rm fus}$ of about 40 MW and $Q = 1$ [3]. The project [3] features a symmetric double-null magnetic configuration and, correspondingly, two divertors that are up-down symmetric. The conceptual design of the divertor and first wall for DEMO-FNS was published elsewhere [8]. Simulations of thermal loads onto the first wall and divertor targets suggested a possibility to distribute the heat loads making them less than 10 MW m^{-2}. Evaluations of sputtering and evaporation of plasma-facing materials suggested that lithium may protect the first wall. It was noted that obtaining the divertor detachment is a goal of further modelling. Here we present the initial explorations of the divertor performance that can be expected in such a system.

The SOLPS4.3 code suite developed for ITER [9] is applied for these studies. It includes a 2D multi-fluid model for the charged components of the plasma and a Monte-Carlo model for the neutrals. The fluid equations are solved on a regular, quadrilateral grid aligned with the magnetic flux surfaces. The parallel heat conductivity and viscosity are classical [10] with flux-limit factors introduced to avoid too fast, non-physical transport at low collisionality [11]. The cross-field diffusivity and heat conductivity are taken spatially constant and, following the approach of [12], their values are adjusted to obtain the width of the power scrape-off layer λ_q consistent with the experimental scaling [13]. For the parameters selected, the scaling [13] yields $\lambda_q \cong 1.5$ mm. The usual sheath boundary conditions are set at the targets and the radial decay length of 3 cm at the side edges of the grid. At the core boundary, the total energy and particle fluxes entering the calculation domain from the core plasma are specified, keeping the flux asymmetry 4:1 in favour of the outboard side.

The neutral model used in the calculations includes atoms and molecules with a wide spectrum of reactions between them and the charged particles. These reactions include multistep ionization of neutrals, 3-body recombination of electrons with ions, charge exchange, molecule-activated recombination (MAR), different channels of molecule dissociation, elastic collisions of molecules with plasma ions and neutral-neutral collisions.

For initial explorations, pure D plasma is selected. The impurity effect is mimicked by reducing the power input to the edge P_{SOL}. Fixing the values of P_{fus} = 40 MW and Q=1 and assuming that ~ 40% of the total heating power is radiated from the core, one gets $P_{SOL} \cong 30$ MW. Assuming that 40% of P_{SOL} is radiated by impurities in the edge, we take the effective value of P_{SOL} = 18 MW in our calculations. Since the scope of the first studies does not include the helium removal issues, the pumping and fueling are not modeled explicitly – the "closed box" approximation is used instead, where the total particle content in the edge is the input parameter and the ion outflow from the core balances the neutral influx there.

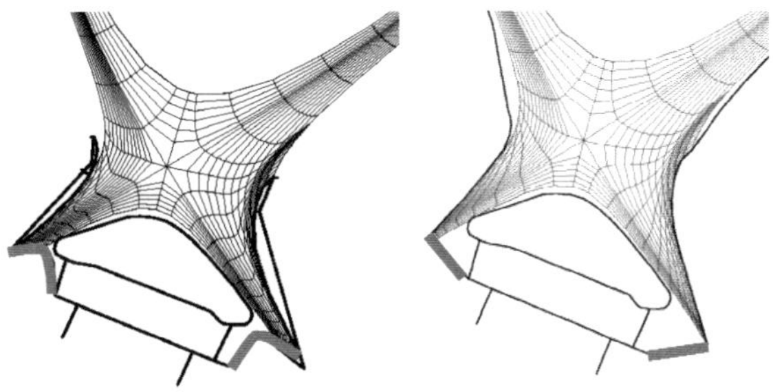

Figure 1. Divertor geometry used in the calculations: the "closed" (left) and "open" (right) versions. The "dome" is introduced to ensure neutral compression. The dome supporting structures are taken into account by introducing semi-transparent surfaces beneath the dome. The top divertor is the mirror image of the bottom one.

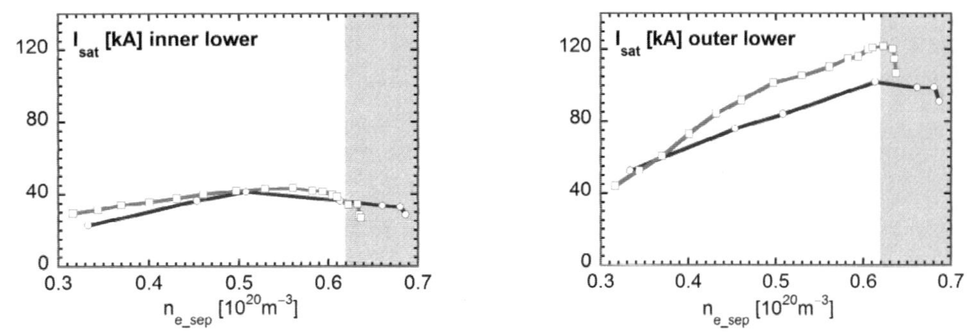

Figure 2. Target-integrated I_{sat} value as a function of the plasma density at the separatrix for the two versions of divertor geometry on the inner (left) and outer (right) lower targets. Here and in the figures below, the blue line corresponds to the closed and the red one to the open divertors.

Fig. 1 shows the divertor geometry used in the calculations. The symmetric double-null divertor is equipped with a "dome" – a solid structure located in the private flux region (PFR) in order to compress the neutrals there. The machine parameters considered are R = 3.2 m, a = 1 m, B_t =5 T, I_p =

5 MA. To take into account the dome-supporting structures, semi-transparent surfaces are introduced between the dome and the divertor bottom. We compare here two versions of the divertor geometry, Fig. 1, differing by the openness of the recycling zones towards the PFR, which should be important for the detachment quality [14].

Fig. 2 shows the dependence of the saturation current integrated over the target, I_{sat}, on the edge plasma density. The I_{sat} roll-over that is the principal signature of detachment is similar in both the geometry versions, occurring at lower density on the inner target, as usual. Following the criterion adopted in ITER modelling [15], the plasma density at the separatrix should be limited from above by the value $n_{e\_sep} \sim 6\ 10^{19}\ \mathrm{m}^{-3}$ in order to avoid deterioration of the core confinement (note however that the correlation of this deterioration with the I_{sat} roll-over is inferred from the single-null experiments [16]). Anyway, this limitation is not severe.

The peak power loading of the outer (most loaded) target is shown in Fig. 3. For both the geometries considered, it increases along with reduction of n_{e_sep} and this could set the lower limit for n_{e_sep}. From Fig. 3, this limitation does not look severe, especially for the open divertor geometry. Recall, however, that we assume here that nearly 40% of the input power is radiated by impurities in the divertors and this may be not the case at low density. The separatrix temperature is also reasonable, $T_e \sim 100$ eV and T_i somewhat higher. Fig. 3 shows also the peak electron temperature at the outer target. One can see that this value is rather high even for the open geometry. This can be a concern from the viewpoint of physical sputtering of the targets since the impinging ions are accelerated in the sheath potential that can reach the values of about $3T_e$. The profiles of T_e at the outer target and of the power flux onto it are shown in Fig. 4. In the open divertor configuration, the temperature at the position of the flux maximum is low, whereas in the closed divertor case, their locations coincide, thus increasing the erosion concern. In the open divertor, the neutrals penetrate higher and attack the divertor plasma from the PFR side, causing preferential reduction of the heat flux along the separatrix [12].

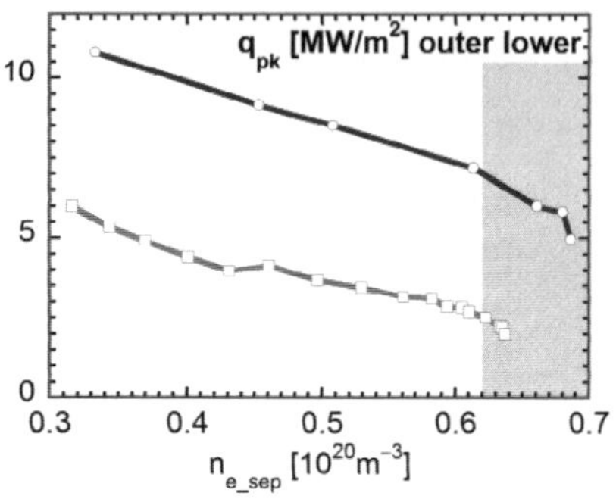

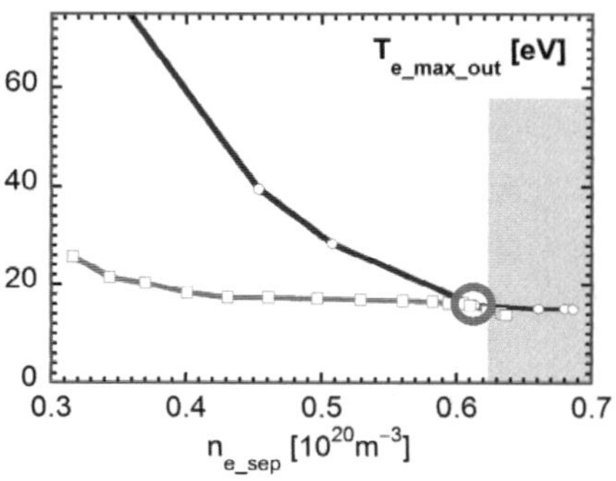

Figure 3. The peak power loading (left) and maximum electron temperature (right) at the outer target vs. separatrix density for the two versions of divertor geometry. The red circle marks the points used for comparing the profiles in Fig. 4.

Fig. 5 shows the pressure in the PFR where the pumping ducts should be located. Here we see more than two orders of magnitude difference between the closed and open divertor options, which means the same difference in the pumping speed required to remove the same particle flux. A rough lower estimate of this flux can be obtained from the helium removal requirements. Assuming the fusion power of 40 MW and He enrichment in the PFR similar to that in ITER [15], one obtains the He pressure in the PFR $p_{He\_PFR} \sim 0.01$ Pa, hence, in the open divertor option, the required pumping speeds are $S_{p\_He} \sim 5\ \mathrm{m^3/s}$ and $S_{p\_DT} \sim 10\ \mathrm{m^3/s}$ and the fuel throughput is $\Gamma_{DT} \sim 30$-$100\ \mathrm{Pa{\cdot}m^3/s}$.

Therefore, the first modelling studies of the DEMO-FNS divertor show that the divertor concept proposed does not look unfeasible (provided that some 40% of P_{SOL} can be radiated from the divertors). The divertor configuration open towards the PFR offers better performance, especially in terms of pumping. Further studies are under way, aimed at the analysis of impurity seeding potential for providing the necessary radiation power and optimization of the pumping conditions.

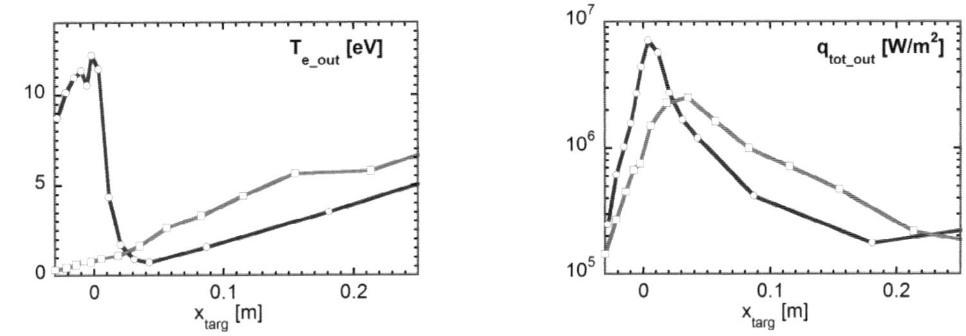

Figure 4. Profiles of electron temperature (left) at and power flux onto the outer target.

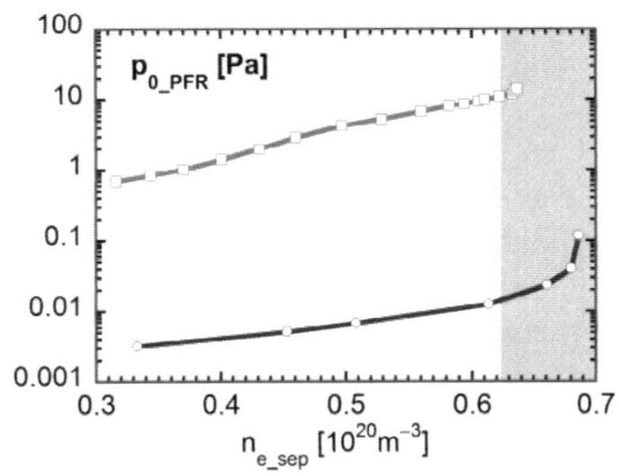

Figure 5. Neutral pressure in the PFR for the open and closed divertor options vs. n_{e_sep}

Acknowledgments
This work was performed using high-performance computing resources of federal center for collective usage at NRC "Kurchatov Institute" (http://computing.kiae.ru/) and Peter the Great Saint-Petersburg Polytechnic University Supercomputing Center (http://www.spbstu.ru).

References

[1] Aymar R et al 2002 *J. Nucl. Mater.* **307–311** 1-9
[2] Jassby D L 1975 *Nucl. Fusion* **15** 453
[3] Kuteev B V et al 2015 *Nucl. Fusion* **55** 073035
[4] Gryaznevich M P et al 2015 *Journal of Physics: Conference Series* **591** 012005
[5] Kuteev B V et al 2011 *Nucl. Fusion* **51** 073013
[6] Chapman I T et al 2015 *Nucl. Fusion* **55** 104008
[7] Sykes A, et al 2017, to appear in *Nucl. Fusion*
[8] Sergeev V Yu et al 2015 *Nucl. Fusion* **55** 123013
[9] Kukushkin A S et al 2011 *Fusion Eng. Des.* **86** 2865–2873
[10] Braginskii S I 1965 *Reviews of Plasma Physics* **1** 205 (New York: Consultants Bureau)
[11] Krall N A and Trivelpiece A W 1973 "Principles of Plasma Physics" McGraw-Hill New York 315-323
[12] Kukushkin A S, et al 2013 *J. Nucl. Mater.* **438** S203–S207
[13] Eich T et al 2013 *Nucl. Fusion* **53** 093031
[14] Kukushkin A S and Pacher H D 2016 *Nucl. Fusion* **56** 126012
[15] Pacher H D et al 2015 *J. Nucl. Mater.* **463** 591-595

[16] Huber A et al 2013 J. *Nucl. Mater.* **438** S139.

IOP Conf. Series: Journal of Physics: Conf. Series **907** (2017) 012013 doi:10.1088/1742-6596/907/1/012013

Globus-M2 NBI upgrade and updated NBCD results obtained on spherical tokamak Globus-M

P B Shchegolev[1,4], V B Minaev [1], A Yu Telnova [1], N N Bakharev [1], P R Goncharov[2], V K Gusev [1], G S Kurskiev [1], I V Miroshnikov [1], M I Patrov [1], Yu V Petrov [1], N V Sakharov[1], I V Shikhovtsev[3] and S Yu Tolstyakov [1]

[1] Ioffe Institute, RAS, ul. Politekhnicheskaya 26, St. Petersburg, 194021, Russia
[2] Peter the Great St. Petersburg Polytechnic University, ul. Politekhnicheskaya 29, St. Petersburg, 195251, Russia
[3] Budker Institute of Nuclear Physics, SB RAS, ul. akademika Lavrentieva 11, Novosibirsk, 630090, Russia

[4] E-mail: peter_shchegolev@mail.ioffe.ru

Abstract. The latest results are presented from NBCD experiments at the Globus-M spherical tokamak. The Globus-M2 NBI upgrade program is discussed, in particular, the analysis of the injection geometry and choice of the optimal experiment layout. The results from numerical simulations of plasma heating and current drive by two neutral beams in the Globus-M2 shots are also analysed.

Introduction

The main problem of a tokamak-reactor which uses an inductor to create toroidal plasma current is its inherently cyclic pulsed operation. The reactor operation should be interrupted cyclically because the magnetization reversal is needed. To provide the continuous operation, the plasma current should be maintained by non-inductive methods. The development of appropriate techniques is a major step in designing a fusion neutron source based on tokamak. In fact, non-inductive current drive techniques are very important for spherical tokamaks where there is very little space for inductor winding. Neutral beam injection (NBI) is a popular and well reproducible technique for current drive in tokamak plasma. That is why one of the main goals of the Globus-M and the future Globus-M2 experiments is the study of the neutral beam current drive (NBCD).

Experimental setup and main research techniques

Neutral beam injection system [1] of the Globus-M spherical tokamak [2] can deliver 1 MW of the additional heating power to the plasma by means of deuterium/hydrogen beam with the energy of up to 30 keV. Experiments were performed in tokamak plasma at two different toroidal magnetic fields of $B_{tor} = 0.4$ and 0.5 T and a plasma current of $I_p = 0.17$ MA. Beams were injected in the equatorial plane of the torus in the tangential direction along the plasma current, the impact parameter being 0.33 m. Hydrogen and deuterium were used as working gases for both the target plasma and injected beam. The electron temperature and density profiles were measured using the Thomson scattering diagnostics. The ion temperature and the losses of fast particles were calculated using data on the fluxes of atoms produced during charge exchange processes and measured by the neutral particle analyzers AKORD-12 and AKORD-24M. The charge exchange recombination spectroscopy [3] was used to measure the ion temperature.

Main results on NBCD

As of today, some progress in the NBCD studies was achieved at the Globus-M facility [4–6]. Non-inductive current drive was observed at the stabilized total plasma current, the value of NBCD being determined from the loop voltage drop. The deuterium beam (0.5 MW, 28 keV) was injected into the deuterium plasma (170 kA) during the stationary stage of the discharge (shot #34275). For the first time, in this shot the radial profiles of the current drive produced by the injection of high-energy atomic beam were calculated basing on the results of [7], the electron shielding current being taken into account (see [8]). The input parameters for the calculations were: the plasma effective charge (Z_{eff} = 2.5), the electron density and temperature profiles, the injection geometry, neutral beam parameters and two different distributions of neutral atoms over the plasma column cross section assumed to be the top and bottom limits (see Figure 1a; ρ is the magnetic surface index). The magnetic configuration was calculated using data on the shapes of closed magnetic surfaces obtained by means of the EFIT code [9].

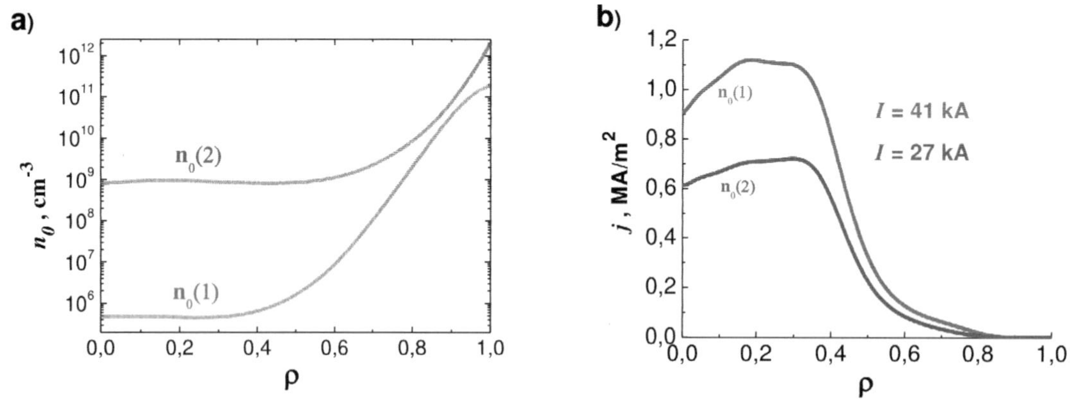

Figure 1. (a) Two different distributions of neutral atoms and (b) radial profiles of the current drive produced by the injection of high-energy atomic beam, the electron shielding current being taken into account.

At first, we calculated the source functions of particles on each magnetic surface for the three energy components of the injected beam taking into account the losses of fast particles. The total cross section of the electron losses for the beam atoms is composed of the cross sections for charge exchange processes of the beam atoms by plasma particles and cross sections for the ionization by the electron and ion impacts. Then, the kinetic equation is solved with allowance for the source functions obtained in the first step of calculations. Further, after calculating the deuteron distribution function, we calculated the fast ion current density profile. The calculation algorithm is described in detail in [7]. In the third step, the shielding effect of electrons is calculated according to [8]. As a result, we obtain the current density profiles for the fast ions produced by NBI for the two different profiles of neutrals' density (shown in Figure 1b) calculated with allowance for the electron shielding current.

In the last series of experiments with NBI on Globus-M, we studied the effect of the increase in the toroidal magnetic field on NBCD (the toroidal magnetic field B_t was increased up to 0.5 T). In these experiments, the 0.95 MW hydrogen beams with particle energy of 28 keV were injected into deuterium plasmas (I_p = 170 kA, $\langle n_e \rangle = 3 \times 10^{19}$ m^{-3}) during the stationary discharge stage at different toroidal magnetic fields. The poloidal magnetic flux consumption caused by NBCD is shown in figure 2. It is clear that the increase in the toroidal magnetic field by 20% (from 0.4 to 0.5 T) results in the lower consumption of the volt-second capacity of the inductor (up to 10%) which is caused by the additional toroidal current driven by the neutral beam. As the toroidal magnetic field grows, the losses of fast particles produced by high-energy atomic beam decrease, and fast ion confinement in plasma improves, resulting in the increase in the NBCD efficiency. Ultimately, we expect to achieve the significant progress in the NBCD studies at the Globus-M2 tokamak, in which the toroidal magnetic field is planned to be increased up to 1 T and a new high-energy neutral beam injector will be installed.

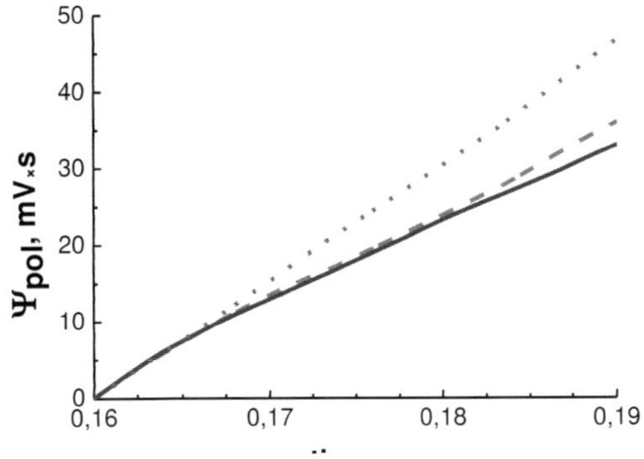

Figure 2. Poloidal magnetic flux consumption for NB heated #36932 (blue solid, B_t = 0.5 T); #37017 (red dash, B_t = 0.4 T) and ohmic #36902 (green dot, B_t = 0.5 T) discharges.

Modernization of the Globus-M NBI system

The upgrade program for the Globus-M tokamak is aimed to increase the toroidal magnetic field and plasma current up to 1 T and 500 kA, respectively; therefore, the plasma density will also rise. To provide the efficient penetration of the fast beam atoms into the core of the plasma column, their energy has to be increased. The Globus-M NBI system upgrade program involves improvement of the installed neutral beam injector and installation of the additional neutral beam injector.

Currently, the production of a new ion source for the already installed injector which will be capable of generating the 1-MW-power neutral beam with energy of particles of up to 40 keV is nearly completed. To support its operations, a step-up autotransformer which allowed for the increase in accelerating electrode voltage to 40 kV was added into the high-voltage power supply circuit.

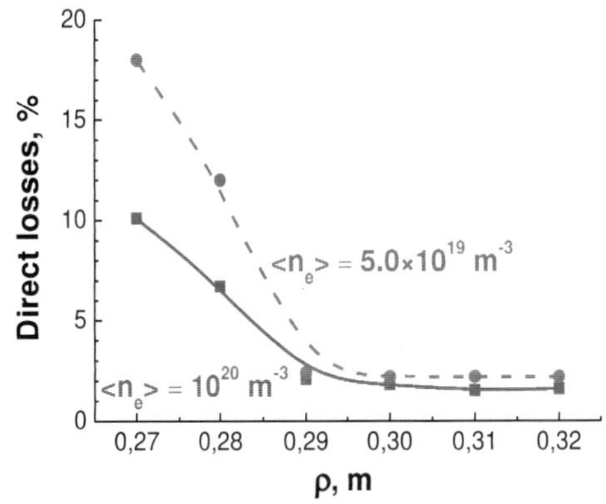

Figure 3. Direct losses of fast particles as a function of the accessible impact parameter (B_{tor} = 0.7 T, I_p = 0.3 MA, E_{beam} = 40 keV).

The installation of the new injector required the analysis of injection geometry options, which served as the foundation for the experiment on the introduction of the second atomic beam in tokamak. Both atomic beams from two injectors were forwarded along the plasma current (co-injection) to accommodate design constraints. The optimal geometry of injection was chosen basing on the results of simulations of the direct losses of fast particles at the accessible impact parameters for the new injector (0.27–0.32 m) performed using the 3D algorithm [10] for tracing the particle trajectories. At the maximum toroidal magnetic field B_{tor} = 1 T and current I_p = 0.5 MA accessible in the Globus-M2 facility, the calculated losses were small (<5%) and were within the computational errors; therefore, calculations were performed at the lower magnetic field B_{tor} = 0.7 T and current I_p = 0.3 MA for two volume-averaged plasma densities (5.0×10^{19} m^{-3} and 1.0×10^{20} m^{-3}) at the beam energy of 40 keV. Figure 3 shows the calculated direct losses of fast particles as a function of the accessible impact parameter of the new injector. Basing on this analysis and taking into account the design constraints, the impact parameter of the second injector was set to be 0.30 m.

Modeling of heating and current drive created by two neutral beams for Globus-M2 discharges

The ASTRA code [11] was used to simulate the main plasma parameters for the Globus-M2 discharges (I_p = 0.5 MA, B_{tor} = 1 T) with the injection of two atomic beams (the first and the second beams have power of 0.9 MW and energy of 40 and 50 keV, respectively); whilst the modeling of fast particle losses was carried out using the fast ion tracking algorithm [10]. Input parameters for the calculation included the electron density profile (volume averaged plasma density 1.0×10^{20} m^{-3}), plasma effective charge (Z_{eff} = 2), parameters of two atomic beams and injection geometry. To calculate the ion density, we assumed carbon to be the main impurity. Energy confinement time τ_E was computed using the model, then compared with those predicted by the ITER H-mode scaling τ_E^{IPB98} [12] and the Valovic scaling τ_E^{ST} [13]. We note that the confinement time τ_E^{ST} generally exceeded the calculated τ_E. In the model, the heat balance equations for electron and ion components were solved (using the NCLASS module) under the assumption that ion and electron thermal conductivities were equal, and they both exceeded the neoclassical thermal conductivity. The relation between the energy confinement times was defined to be $\tau_E = 1.3 \times \tau_E^{IPB98}$ (Fig. 4a).

The results of these calculations are shown in Figure 4b. Numerical simulations of the parameters of the Globus-M2 discharges with the injection of two neutral beams predict an increase in the ion and electron temperatures accompanied by the increase in the stored energy, energy confinement time and NBCD efficiency. In addition, calculations show that the total fraction of the current driven non-inductively will amount to approximately 40-45%.

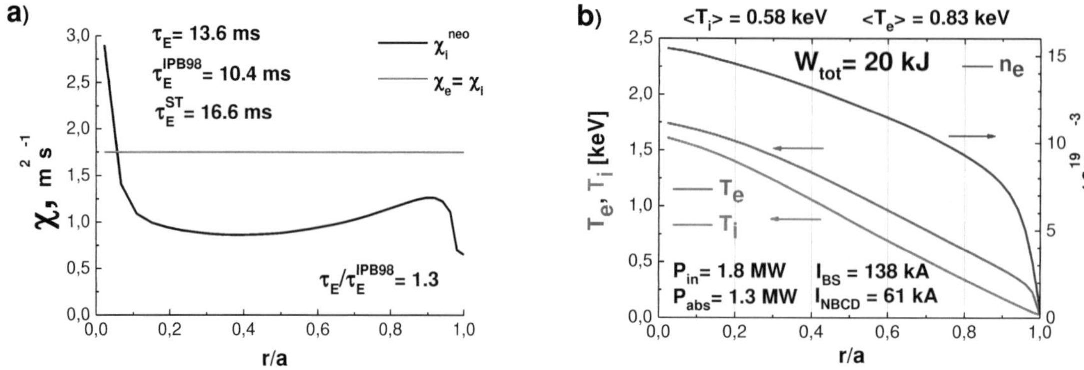

Figure 4. (a) Ion and electron heat diffusivities as compared to the neoclassical heat diffusivity calculated using the NCLASS module; (b) simulation results of current drive and heating by two neutral beams (50 keV and 40 keV); P_{in} is input power, P_{abs} is absorbed power, and W_{tot} is total stored thermal energy.

Conclusions

For the first time, we calculated the current density profiles for the fast ions produced by the injection of high-energy atomic beam into the Globus-M plasma with allowance for the electron shielding current.

The effect of the increase in the toroidal magnetic field (from 0.4 to 0.5 T) on the NBCD was studied. With the toroidal magnetic field growing by 20%, the poloidal magnetic flux consumption decreased (by nearly 10%), which was caused by the NBCD efficiency growth.

The production of the new ion source of the installed injector capable of generating 1 MW NB with the energy up to 40 keV is nearly completed. For the new injector, optimal experimental layout was chosen, and its impact parameter was set to be 0.3 m.

Numerical simulations of the Globus-M2 discharges with the injection of two neutral beams predict an increase in the ion and electron temperatures accompanied by an increase in the stored energy, energy confinement time and NBCD efficiency. In addition, calculations show that the total fraction of the current driven non-inductively will amount to approximately 40–45%.

Acknowledgments
This work was supported by the Ministry of Education and Science of the Russian Federation (project no. 14.W01.16.5846-MK) and the Russian Foundation for Basic Research (project no. 16-32-00454 mol_a).

References

[1] Gusev V K et al 2007 Technical Physics **52** pp 1127-43
[2] Gusev V K et al 1999 Technical Physics **44** pp 1054-57
[3] Avdeeva G F et al 2016 Journal of Physics: Conference Series **666** 012002
[4] Minaev V B et al 2015 Proc. 42rd EPS Conf. on Plasma Phys. (Portugal) 122445
[5] Shchegolev P B et al 2015 Technical Physics **60** pp 1321-25
[6] Shchegolev P B et al 2016 *Proc. 43rd EPS Conf. on Plasma Phys. (Leuven)* 125175
[7] Goncharov P R et al 2011 Nuclear Fusion **51** 103042
[8] Hu Y J Hu Y M Lin-Liu Y R 2012 *Phys. Plasmas* **19** 034505
[9] Gusev V K et al 2006 Technical Physics **51** pp 987-94
[10] Bakharev N N et al 2015 Nuclear Fusion **55** 043023
[11] Pereverzev G V et al 1991 Report IPP 5/42
[12] ITER Physics Basis 1999 Nuclear Fusion **39** p 217
[13] Valovic M et al 2009 Nuclear Fusion **49** 075016

IOP Conf. Series: Journal of Physics: Conf. Series **907** (2017) 012014 doi:10.1088/1742-6596/907/1/012014

Effect of increase in the toroidal magnetic field on plasma heating and confinement in the Globus-M tokamak discharges with neutral beam injection

A Yu Telnova[1], V B Minaev, P B Shchegolev, N N Bakharev, V K Gusev, G S Kurskiev, Yu V Petrov, N V Sakharov

Ioffe Institute, 194021, St. Petersburg, Politehnicheskaya st, 26

[1] E-mail: anna.telnova@mail.ioffe.ru

Abstract. In the course of modernization of the Globus-M spherical tokamak, the experiments at the increased toroidal magnetic field (from 0.4 to 0.5 T) and with the additional heating by neutral beam injection were carried out. The numerical simulations were performed using the ASTRA transport code. A considerable increase in the absorbed neutral beam power as well as in the energy lifetime and total energy content was achieved due to the increase in the toroidal magnetic field and plasma current.

1. Introduction

The authors have more than once noted that the low toroidal magnetic field (0.4 T) in the Globus-M spherical tokamak [4] is one of the main reasons for limiting the further improvement of plasma parameters [1–3]. The low magnetic field also reduces the possibilities of using the additional plasma heating and current drive techniques. In the upgraded Globus-M2 facility, the toroidal magnetic field will be increased up to 1.0 T [5]. A new electromagnetic system and power supplies were developed and manufactured.

During the last experimental campaign, all new power supplies were tested in experiments at the Globus-M tokamak. They were connected to the coils of the existing electromagnetic system. The variation range of feeding currents was increased by 25%. Under these conditions, a series of experiments on plasma heating by the neutral beam injection (NBI heating) was carried out. These experiments were aimed to study the effect of the toroidal magnetic field and plasma current on the confinement of fast ions appearing in the course of the neutral beam injection as well as on the plasma confinement.

2. The results of experiments on plasma NBI heating

In the experiments, the toroidal magnetic field and plasma current varied in the ranges of 0.4–0.5 T and 200–250 kA, respectively. The deuterium beam (26 keV, 0.7 MW) was injected into deuterium plasma in the stationary stage of the discharge (at 160 ms). The beam duration (40 ms) manyfold exceeded the typical energy lifetime τ_E. During the beam injection phase, the mean electron density was maintained at the level of $3–4 \cdot 10^{19}$ m^{-3}.

Figure 1 shows time evolution of the plasma parameters for three shots with different toroidal magnetic fields and plasma currents. The shot #36594 is a typical of the Globus-M shot with the usual mean density, plasma current of about 200 kA and the toroidal magnetic field of 0.4 T on the axis. In the shot #36610, the toroidal magnetic field was increased up to 0.5 T, the plasma current being the same. And in the shot #36618, the plasma current was increased up to 250 kA and the toroidal magnetic field remained 0.5 T. The electron temperature and density profiles were measured using the Thomson scattering diagnostics. The measured radial profiles of the electron temperature T_{e_exp} for these three shots are shown in Figure 2 by black dots. The mean chord density was monitored by a microwave interferometer. The ion component of the plasma was studied by means of two neutral particle analyzers, which measured the fluxes of neutral atoms coming from plasma after the processes of charge exchange in collisions with ions. The analyzers were oriented in tangential and transversal to the beam line directions. These data were used to estimate the fast particle losses, the ion temperature, ion composition and the absorbed beam power. The total energy content of plasma was estimated from the diamagnetic measurements using the set of loops.

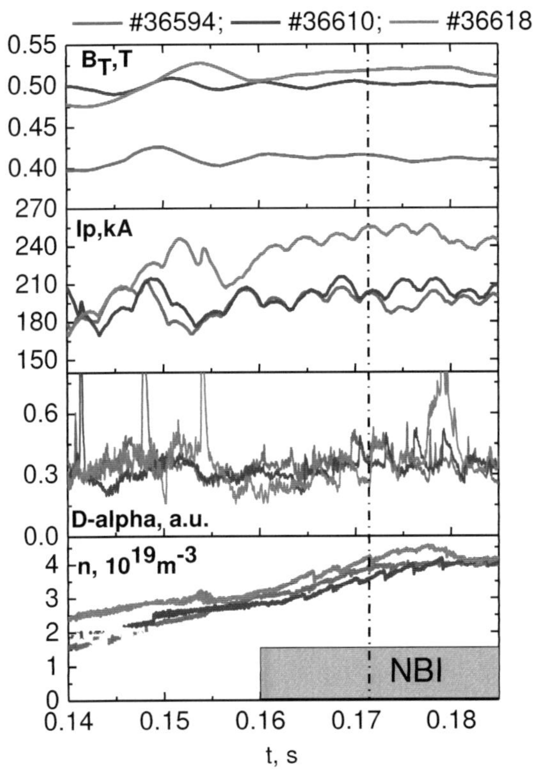

Figure 1. Time evolution of the plasma parameters in the shots with different toroidal magnetic fields and plasma currents: green, blue and red curves correspond to the shots #36954 (0.4 T, 200 kA), #36610 (0.5 T, 200 kA) and #36618 (0.5 T, 250 kA), respectively. From top to bottom: toroidal magnetic field, plasma current, D-alpha line emission intensity, volume averaged plasma density.

3. Description of the transport code

The ASTRA transport code [6] was used for processing of the experimental results. In simulations, we studied the thermal diffusivity coefficients, total plasma energy content and energy lifetime as functions of the toroidal magnetic field and plasma current.

The modeling was carried out in the quasi-stationary stage of discharge (~170 ms) under the following assumptions. Assuming the transport coefficients to be neoclassical, we calculated the ion temperature profiles using the NCLASS transport code [7]. As the boundary conditions for solving the balance equation using the ASTRA code, we used the parameters of the last closed magnetic surface, recovered using the EFIT code [8]. The effective plasma charge was assumed to be constant along the minor radius. The Z_{eff} value was determined by comparing the estimated and measured loop voltages. Carbon was assumed to be the main impurity. The absorbed power of the neutral beam was calculated using a standard block of the ASTRA code [9]. The additional corrections to the absorbed power were made in order to take into account the fast particle losses from the first orbits. These losses were determined using the 3D modeling algorithm [3].

4. Discussion of the simulation results

The simulation results of the electron and ion temperatures as compared to the experimental data are shown in the Fig. 2. The figure demonstrates the considerable increase in the electron and ion temperatures with increasing both the toroidal magnetic field and plasma current. In each of the shots considered, the effective plasma charge was about 2.4–2.5. Table 1 shows the energy contents of plasma W_{ASTRA} and W_{diam} calculated using the ASTRA code and from data of the diamagnetic measurements, as well as the energy lifetimes $\tau_{E\ ASTRA}$ and $\tau_{E\ IPB98(y,2)}$ calculated using the ASTRA code and the scaling IPB98(y,2) [11]. Data on the total absorbed power, the absorbed power of the beam with allowance for the losses and the other integral characteristics of the discharges are also given in Table 1.

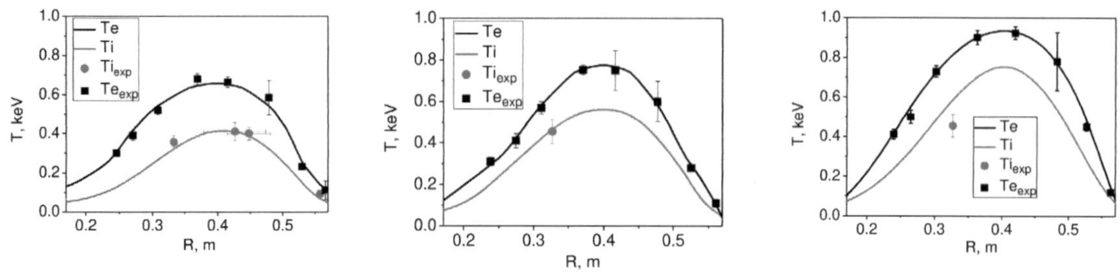

(a) $B_T = 0.4$ T $I_p = 200$ kA #36954 (b) $B_T = 0.5$ T, $I_p = 200$ kA #36610 (c) $B_T = 0.5$ T, $I_p = 250$ kA #36618

Figure 2. Electron (black) and ion (red) temperature profiles calculated and measured at 170th ms (lines and dots, respectively) for three shots with different toroidal magnetic fields and plasma currents.

Increase in the toroidal magnetic field by 25% at a fixed plasma current of 0.2 MA resulted in the rise of the total energy content by 30% which is confirmed by the diamagnetic measurements. In this case, the total absorbed power didn't considerably change because the efficiency of fast ions confinement which determines power of additional heating better depends on the plasma current but not on the toroidal magnetic field [3]. Therefore, an increase in the energy content is a consequence of the improved confinement of thermal energy and exactly the electron energy confinement is improved at that (see Fig. 3 and Table 1). An increase in the plasma current up to 0.25 MA at the fixed magnetic field $B_T = 0.5$ T results in the rise of the absorbed power by 20%, the total energy content and energy lifetime being increased by 40 and 20%, respectively (as compared to the typical discharge). Heat diffusivities for electrons and ions are presented in Figure 3. The electron heat diffusivity decreases both with the rising current I_p and magnetic field B_T. The dependence of the confinement time on the plasma current and magnetic field was $\tau_E \sim B_T^{0.8\pm0.1} I_p^{0.8\pm0.1}$.

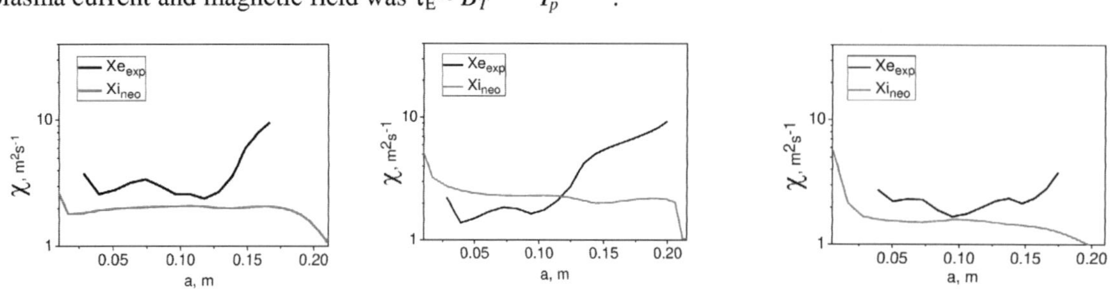

(a) $B_T = 0.4$ T $I_p = 200$ kA #36954 (b) $B_T = 0.5$ T, $I_p = 200$ kA #36610 (c) $B_T = 0.5$ T, $I_p = 250$ kA #36618

Figure 3. Heat diffusivities for electrons χ_e (black) and ions χ_i (red)

Table 1. Comparison between the calculated and measured plasma parameters for the different magnetic fields and plasma currents

Discharge	#36594	#36610	#36618
Toroidal magnetic field, B_T, T	0.4	0.5	0.5
Plasma current, I_p, kA	200	200	250
Z_{eff}	2.4	2.5	2.5
Full absorbed power, P_{abs}, MW	0.527	0.543	0.667
NBI absorbed power, $P_{nbi\ abs}$,MW	0.285	0.32	0.389
W_{diam}, kJ	1.7	2.38	3.42
W_{fast}, kJ	0.24	0.36	0.51
W_{ASTRA}, kJ	1.56	2.02	3.11
$\tau_{E\ ASTRA}$, ms	3	3.7	4.7
$\tau_{E\ IPB98(y,2)}$, ms	4.3	4.6	5.2
$H_{IPB98(y,2)} = \tau_{ASTRA}/\tau_{IPB98}$	0.7	0.8	0.9

The studies carried out at the MAST spherical tokamak [12] have shown that the dependence of the energy lifetime on the toroidal magnetic field is much more strong than that given in the scaling IPB98(y, 2), and at the same time the corresponding dependence on the current is linear. The similar dependence on the magnetic field was observed at the NSTX tokamak, however, the dependence on the plasma current is linear under conditions of boronization of the vacuum chamber walls performed after their cleaning by the glow discharge [13].

The experiments on the additional NBI plasma heating carried out at the Globus-M tokamak demonstrated that the dependences of energy lifetime on the toroidal magnetic field and on plasma current are close to linear at a moderate mean density of $\sim 4 \cdot 10^{19}$ m^{-3}. This result is hopeful to expect the considerable improvement of the plasma parameters in the Globus-M2 spherical tokamak which will operate at the higher current I_p and toroidal magnetic field B_T.

Acknowledgments
This work was supported in part by the Russian Science Foundation, project no. 16-32-00454 mol_a (P.B. Shchegolev) and the Russian Federation Presidential Grant No. MK-5846.2016 (Kurskiev G.S. and Telnova A.Yu.).

References

[1] Gusev V K et al 2013 *Nucl. Fusion* **53** 093013
[2] Gusev V K et al 2015 *Nucl. Fusion* **55** 104016
[3] Bakharev N N et al 2015 *Nucl. Fusion* **55** 043023
[4] Gusev V K et al 1999 *Tech. Phys.* **44** 1054
[5] Minaev V B et al 2017 *Nucl. Fusion* **57** *to be published*
[6] Pereverzev G and Yushmanov P N 2002 *Max-Plank IPP Report* 5/98
[7] Houlberg W A et al 1997 *Phys. Plasmas* **4** 3230
[8] Lao L L et al 1985 *Nucl. Fusion* **25** 1611
[9] A Polevoi, H Shirai and T Takizuka JAERI -Data/Code 97-014 March 1997
[10] Avdeeva G F et al 2016 *Proc. 43d EPS Conf. on Plasma Physics (Belgium, Leuven, 4 - 8 July, 2016).* P1.064
[11] ITER Physics Basis 1999 *Nucl. Fusion* **39** 2175, 2007 *Nucl Fusion* **47** S18
[12] Valovic M et al 2009 *Nucl. Fusion* **49** 075016
[13] Kaye S M et al 2006 *Nucl. Fusion* **46** 848–57
[14] Kurskiev G.S. et. al 2017 *Plasma Phys. Control. Fusion* **59** 045010

Peculiar properties of internal transport barrier formation near the q = 1 and q = 2 surfaces in tokamaks

S V Neudatchin[1], D A Shelukhin and N A Mustafin

NRC Kurchatov Institute, Russian Federation, Moscow, 123182, Kurchatov sq 1

[1] E-mail: sneudat@yandex.ru

Abstract. A new type of internal transport barrier (ITB) was found in the T-10 tokamak plasma for the first time. It is created by sawtooth oscillations which are almost damped by off-axis electron cyclotron resonance heating and electron cyclotron current drive (ECRH and ECCD). During the ITB formation, the electron temperature, T_e, at a relative radius $r/a \leq 0.45$ rapidly rises to a new quasi-steady state, while the turbulence level, which is measured at the ITB, falls below its pre-crash value, and the turbulence spectrum shrinks in a wide range of frequencies. The ITB appears near the q = 1 magnetic surface and exists for approximately 50–70% of the plasma energy confinement time. During this process, accumulation of impurities was not observed. An abrupt and non-local reduction of the electron heat flux (ITB-event) around the q=1 surface (within 1 ms in the spatial zone with a relative radius $0.2 < r/a < 0.5$) was observed for the first time for the case with a gas puffing cut-off in OH T-10 discharges with an increasing high density. In contrast to the ITB that forms during the ECR heating, in this case, impurities begin to accumulate simultaneously with the ITB event. Analysis of the D-III-D experiments with a weak reverse magnetic shear shows that the appearance of a $q_{min} = 2$ surface leads to this ITB event in a very wide spatial region (the relative radius of the plasma column is $0 < r/a < 0.7$).

1. Introduction

The important role of low-order rational magnetic surfaces with a safety factor of q = 1, 1.5, 2, 2.5, 3… in the formation of internal transport barriers (ITBs) has been demonstrated earlier both in the plasmas of T-10 tokamak (major radius R = 150 cm, minor radius a = 30 cm) [1–6] and in the plasmas of many other tokamaks (e.g., see review [7]). In T-10, ITB has been recognized by analyzing the slow heat pulse propagation (HPP), which is induced by the onset of central electron cyclotron heating (ECRH) in a sawtooth-free plasma created via off-axis ECRH at the first and second harmonics [1–2]. The ITB forms near the q = 1 surface. The abrupt non-local reduction of transport (ITB-event) in the central part of plasma column often occurs together with the appearance [3] (or slightly before [4]) of the q = 1 surface in T-10 sawtooth-free plasmas. The role of magnetic shear has been demonstrated in T-10 experiments with co/contra ECCD (electron cyclotron current drive) at the center of plasma column [5–6].

The paper shows new peculiarities of the local and non-local processes of ITB formation in the plasmas of T-10 and D-III-D tokamaks. The results of chapter 2 have been reported at the 25-th FEC conference [8], and the full paper has been presented at the International Zvenigorod Conference on Plasma Physics and Controlled Fusion in 2017.

IOP Conf. Series: Journal of Physics: Conf. Series **907** (2017) 012015 doi:10.1088/1742-6596/907/1/012015

2. ITB created by sawtooth oscillations is almost damped by off-axis ECRH/ECCD in the T-10 tokamak

A new phenomenon is observed in the ECR heated plasma in the stage after sawtooth oscillations with a period exceeding 30 ms (their usual period is 4–7 ms) become almost damped by off-axis ECRH/ECCD [8]. Figure 1 shows the unusual evolution of T_e in the T-10 tokamak shot with a 100-ms period of sawtooth oscillations. Sawtooth oscillations create the rise of T_e at the radius r = 16 cm during the fast MHD phase of the current crash in the tokamak, and then the initial rise of temperature disappears within 1–1.5 ms (the limiter is placed at 30 cm). After the current crash, the phase inversion radius r_s and the q = 1 surface are located near the radius r = 12 cm. Later, the value of T_e (at the radius r = 12 cm) increases by 12% during 1-2 ms at the radius r = 12 cm, and the value of grad (T_e) increases by 40% between r = 12.5 and 16 cm. The heat pulse does not propagate outside the plasma column during the ITB existence (~20 ms). The value of electron thermal conductivity coefficient χ_e becomes 2.5 times lower compared with its value determined using the L-mode scaling on the T-10 tokamak. Figure 2 shows the evolution of turbulence during similar crashes in the T-10 tokamak shot 66049 (I_p = 250 kA, B_t = 2.15 T, P = 1 MW, $n_{\text{line av}} = 3\ 10^{19}\ \text{m}^{-3}$, I_{ECCD} = 6 kA), which is measured using a three-wave heterodyne correlation reflectometer (see details in [9]). The reflectometer data shows an enhanced turbulence level during 1–2 ms after the crash at frequencies 40-120 kHz. Later, during the ITB formation slightly outside the q = 1 surface (the outer edge of ITB lies at r ≈ 16 cm), the turbulence level falls slightly below its pre-crash stationary value, and the spectrum of the turbulence shrinks.

The simulations of sawtooth behavior preformed using the transport code ASTRA [10] for the T-10 tokamak shot no. 63446 are shown in Fig. 1. The quantities and profiles of the absorbed EC power and driven current were calculated by solving the Fokker-Plank equation using the OGRAY code [11]. The calculations using the Kadomtsev reconnection model [12] supplied with the critical shear criterion [13] reasonably describe the sawtooth period. This is the case of the most effective sawtooth damping, which occurs at ECCD slightly outside of q = 1 [14]. Figure 3 shows the evolution of the q profile, which is calculated between the current crashes, which are shown in Fig. 1. The EC current is generated in the narrow ~1.5-cm-wide zone (the poloidal angle of the mirror is 8°). The region with a reversed magnetic shear appears soon after the crash in the same region and survives for a long time.

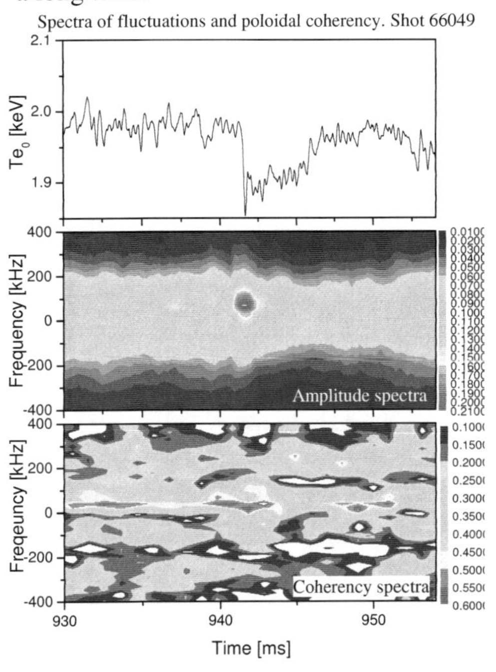

Figure 1. Timetraces of T_e in the T-10 tokamak shot no. 63446 (I_p = 220 kA, B_t = 2.15 T, P = 0.8 MW, $n_{e\ \text{line av}} = 1.8 \cdot 10^{19}\ \text{m}^{-3}$, I_{ECCD} = 10 kA). T_e increases at radius r = 12 cm, and the heat pulse does hot propagate outside the plasma column.

IOP Conf. Series: Journal of Physics: Conf. Series **907** (2017) 012015 doi:10.1088/1742-6596/907/1/012015

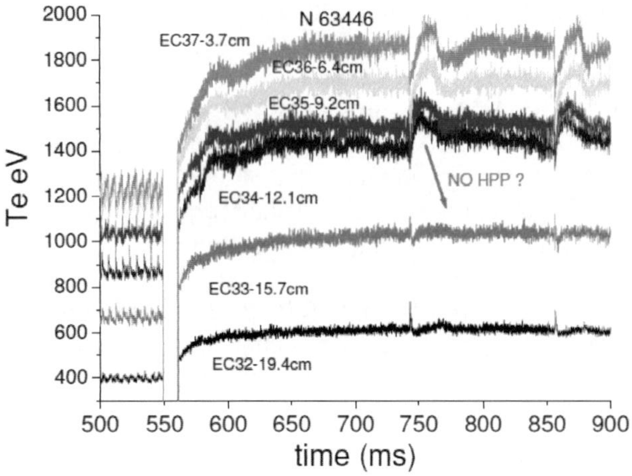

Figure 2. Timetraces of T_e at the center of the plasma column and turbulence spectrum amplitude in a different shot with a weaker ITB.

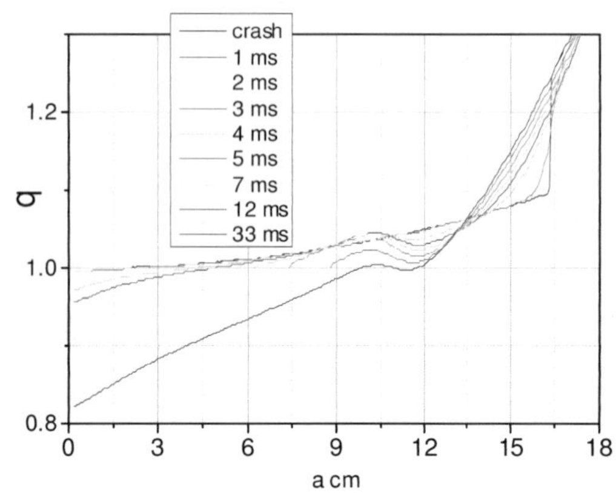

Figure 3. Evolution of q profiles, which are calculated between the crashes that are shown in Fig. 1.

3. Abrupt and non-local reduction of the electron heat flux (ITB-event) around the q = 1 surface after gas puffing is cut-off in the OH regime in T-10 plasmas

The abrupt cut-off of gas puffing in discharges with high-density plasmas (in this case, the electron density n_e at the center of plasma column is equal to $7 \cdot 10^{19}$ m^{-3}) initially creates a reduction of electron density n_e at the plasma edge starting at t = 823 ms. Figure 4 shows the evolution of electron temperature T_e. The temperature increase starts simultaneously with an 8-ms delay in the plasma column zone with a radius 5 cm < r < 15 cm (the inversion radius or q = 1 surface is at the relative radius r/a = 0.25), while the variation of electron density n_e is small at the central part of the plasma column. Since the variation of electron temperature gradient grad T_e is negligible, the following simple expression allows us to calculate the jump of the electron heat flux $\delta\Gamma_e$, where A is the enclosed surface [15]:

$$<\delta\Gamma_e A> = -<\delta\Gamma_e> \{n_e \nabla T_e A(r)\} = -1.5 \left\{ \int_0^{V(r)} n_e <\delta(\partial T_e/\partial t)> dV \right\} \qquad (1)$$

The values of $\delta\Gamma_e$ and χ_e decrease in a wide radial zone (30–40% of the minor radius) and decrease nearly twofold at a relative radius r/a = 0.4–0.5. In contrast to the case reported in chapter 2, impurities begin to accumulate simultaneously with the ITB event. The rise of soft X-ray (SXR) radiation strongly exceeds the rise of T_e since
$(\delta I_{sxr}/dt) / I_{sxr} \approx 6 \, (\delta T_e/dt) / T_e$).

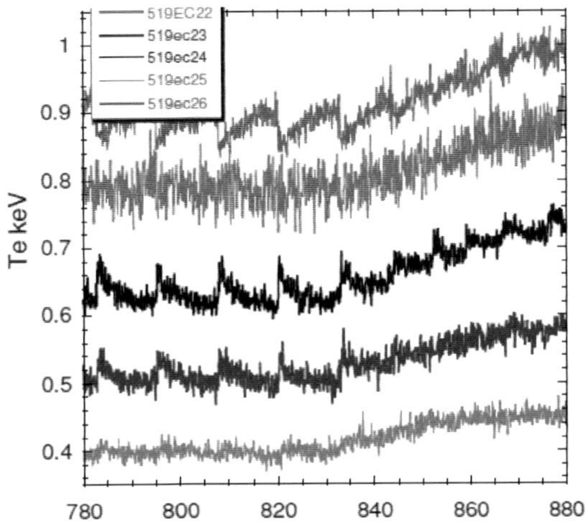

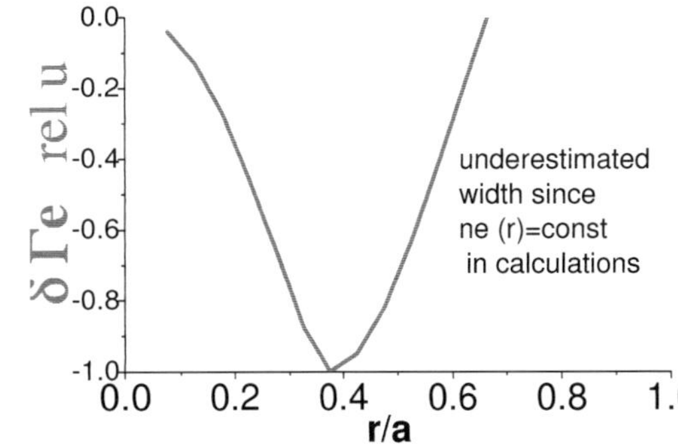

Figure 4. Evolution of electron temperature T_e at r = 5.2, 7.6, 10.1, 12.7, 15.3 cm in the T-10 OH shot 57519.

Figure 5. Profile of $\delta\Gamma_e$ at $q_{min} = 2$ in the D-III-D tokamak.

4. Abrupt and non-local reduction of the electron heat flux (ITB-event) at the appearance of the $q_{min} = 2$ surface in the D-III-D tokamak

The evolution of T_e in weak reverse shear plasmas during the gradual decay of the minimum q value, q_{min}, was presented in [16]. The appearance of $q_{min} = 2$ surface results in the increase of electron temperature T_e inside the magnetic surface q_{min} and electron temperature decay outside this surface. The T_e perturbation profiles after the ITB event, which is presented in [16], were obtained 20 ms after the ITB event. Heat pulse propagation broadens the profile of $\delta\Gamma_e$, which is obtained using equation (1), by averaging over a significant time interval. We use a 5-ms time interval and a constant electron density $n_e(r)$ = const [this reduces spatial width of the heat flux $\delta\Gamma_e$, see Eq. (1)]. The application of expression (1) to the data presented in Fig. 5 [16] brings the appearance of a very wide profile of the electron heat flux perturbation $\delta\Gamma_e$ at a relative plasma column radius $0 < r/a < 0.7$ around the location of q_{min}. The width of $\delta\Gamma_e$ is wider compared with the one calculated in similar JT-60U tokamak experiments with a stronger reverse shear (there, the relative radius is $0.3 < r/a < 0.7$) [17]. Nevertheless, the maximum of the absolute value of the heat flux reduction $\delta\Gamma_e$ lies near the position of q_{min} in both cases.

5. Conclusions

A new type of ITB, which is created by almost damped sawtooth oscillations (the ones with a strongly increased period) and off-axis ECRH/ECCD, is found in the T-10 tokamak for the first time. The electron temperature T_e increases outside the phase inversion radius r_s of the oscillations just after a sawtooth crash. The reflectometer data shows the significantly enhanced turbulence level and spectrum amplitude of the density fluctuations during ~1 ms after the crash in the outside wide spatial zone (in a wide range of frequencies, mainly at 40–120 kHz). The electron temperature T_e decays fast during ~1 ms after the crash. Later, during the ITB formation, the electron temperature T_e increases inside the relative radius r/a = 0.45 (q = 1 surface) to a new steady-state, and the heat pulse does not propagate outside the plasma column during 15–20 ms (50–70% of the plasma energy confinement time). The plasma heat conductivity χ_e value decreases by 2.5 times compared with the L-mode scaling, and the accumulation of impurities is absent. The turbulence level, which is measured at ITB, falls below its pre-crash value, and the turbulence spectrum shrinks in a wide range of frequencies. The calculations, which are performed using the ASTRA/OGRAY codes, show the appearance of the reversed magnetic shear zone near the q = 1 surface soon after the crash. The appearance of the reverse shear appear is a necessary condition for the formation of the ITB.

The abrupt and non-local reduction of the electron heat diffusivity (ITB event) around the q = 1 surface (in the zone 0.2 < r/a < 0.5) in the new case is also reported for the first time. The ITB event occurs after the cut-off of the gas puffing in OH T-10 discharges with a high increasing density. In contrast with the ECRH case described above, impurities begin to accumulate simultaneously with the ITB event.

The analysis of the experiments with a weak reverse shear on the D-III-D tokamak shows that the appearance of the minimum q value surface with q_{min} = 2 leads to the ITB event in a very wide spatial region (0 < r/a < 0.7). A similar event also occurs in JT-60U tokamak plasmas with a stronger reverse shear around the position of q_{min} in a more narrow region (0.3 < r/a < 0.7) [17]. The maximum of the absolute value of the heat flux reduction lies near the position of q_{min} in both cases.

Acknowledgments
The authors are thankful to Drs. N.A. Kirneva, D.A. Kislov, Yu.D. Pavlov and V.A. Vershkov for the fruitful discussions and experiments. The work was supported by the ROSATOM State Corporation Contract H.4x.241.9Б.17.1011.

References
[1] Bagdasarov A, Vasin N L, Neudatchin S V and Savrukhin P V 1991 *Plasma Ph. and Control Nucl. Fus. Res.* (Proc. 15th Int. Conf., Washington, 1990) **1** (Vienna: IAEA) 253
[2] Neudatchin S V., Kislov A. Ya, Krupin V. A. et al 2003 *Nucl. Fusion* **43** 1405
[3] Neudatchin S., Inagaki S., Itoh K., Kislov A., et al 2004 *J. Plasma and Fus. Res. Series* **6** 134
[4] Neudatchin S V., et al 2006 *Proc. 21st IAEA Fusion Energy Conf.* (Chengdo, 2006) EX/P1-8
[5] Razumova K A et al 2000 *Plasma Phys. Control. Fusion* **42** 943
[6] Kirneva N A et al 2000 *27th EPS Conf. on Cont. Fus. and Pl. Phys.* (Budapest, 2000) P2.031
[7] Wolf R C 2003 *Plasma Phys. Control. Fusion* **45** R1
[8] Neudatchin S.V., Shelukhin, D.A. et al. *2014 25th FEC Conf.* (St. Petersburg, 2014) EX/P1-43
[9] Vershkov V.A., Dreval V.V., Soldatov S.V. et al 2005 *Nucl. Fusion* **45** S203
[10] Pereversev G.V. et al 1992 *Report* IAE 5358/6 (Kurchatov Institute, Moscow)
[11] Zvonkov A.V. *et al* 1998 *Plasma Phys. Rep.* **24** 389
[12] Kadomtcev B.B. 1975 *Sov. J. Plasma Phys.* **1** 389
[13] Porcelli F., Boucher D. and Rosenbluth M.N. 1996 *Plasma Phys. Control. Fusion* **38** 2163
[14] Kislov D.A. et al 2007 *Nucl. Fusion* **47** S590 19
[15] Cordey J.G., Muir D.J., Neudatchin S.V. et al 1994 *Plasma Phys. Control. Fusion* **36** A267
[16] Greenfield C.M. et al. 1998 *Plasma Phys. and Control Nucl. Fus. Res.* (Proc. 19th Int. Conf., Kyoto, 1998) vol 1 (Vienna: IAEA) p 258
[17] Neudatchin S V, Takizuka T, Shirai H, et al., 1999 *Plasma Phys. Control. Fusion* **41** L39

Plasma confinement during ECR heating with a volume power density of 3 mW/m^3 at the L-2M stellarator

A I Meshcheryakov[1,6], G M Batanov[1], V D Borzosekov[1,2], S E Grebenshchikov[1], I A Grishina[1], N K Kharchev[1], Yu V Kholnov[1], L V Kolik[1], E M Konchekov[1,2], L M Kovrizhnykh[1], A A Letunov[1], V P Logvinenko[1,5], D V Malakhov[1,2,4], A E Petrov[1,2], K A Sarksyan[1], S V Shchepetov[1], N N Skvortsova[1,2,4], V D Stepakhin[1,2], M A Tereshchenko[1], I Yu Vafin[1] and D G Vasilkov[1,3]

[1] Prokhorov General Physics Institute of the Russian Academy of Sciences, ul. Vaviliva 38, 119991 Moscow, Russia
[2] Pirogov Russian National Research Medical University, ul. Ostrovityanova 1, Moscow, 117997 Russia
[3] Bauman Moscow State Technical University, ul. Vtoraya Baumanskaya 5, Moscow, 105005 Russia
[4] Moscow Terchnological University (MIREA), pr. Vernadskogo 78, Moscow, 119454 Russia
[5] RUDN University, Miklukho-Maklaya str. 6, Moscow, 117198 Russia

[6] E-mail: meshch@flp.gpi.ru

Abstract. The experiments on ECR plasma heating were carried out at the L-2M stellarator at very high volume power density (up to 3.0 MW/m^3). Under these conditions, non-monotonous hollow density profiles were measured. At the maximum heating power of P = 0.75 MW, the concavity in the axial region becomes drastic $n_e(0)/n_e^{max}$ = 0.5. In these experiments, plasma temperature profiles measured in the axial plasma region r/a_p < 0.4 occur to be rather flat. We have also measured two-slope SXR spectra in the range from 2 to 12 keV. The possible reasons for these observed phenomena were discussed. However, with growing ECRH power up to P_{ECRH} = 0.75 MW, we have not observed dramatic deterioration of plasma confinement at the L-2M stellarator. The measured energy lifetime is generally consistent with that determined from the international LHD scaling.

1. Introduction

During several years, the experiments on plasma creation and heating by microwave radiation at the second harmonic of the electron cyclotron frequency were carried out at the L-2M stellarator [1]. A distinctive feature of these experiments is the high volume power density of the ECR plasma heating. Under these conditions, a number of specific features of plasma heating and confinement were revealed at the L-2M stellarator.

The gyrotron complex MIG-3 was used for plasma heating [2]. It consists of two gyrotrons with depressed collector and two quasi-optical channels. One gyrotron operates at a frequency of f = 75 GHz and has the rated microwave power of P = 0.8 MW. The second gyrotron can operate at three frequencies of f = 71.5, 74.8 and 78.2 GHz and has the rated microwave power of P = 0.7 MW. In these experiments, the maximum total power of microwave radiation launched into the stellarator plasma was P = 0.75 MW. In this case, power density of plasma heating amounts up to P/V = 3.0 MW/m^3, where V = 0.25 m^3 is plasma volume. In all experiments with ECR heating, the walls of the vacuum chamber were covered with boron-carbon film in order to reduce impurity content in plasma [3]. In this report we discuss some effects that occur in plasma due to ECR heating under these conditions.

2. Density pump-out effect during ECRH

The experiments on plasma confinement at the L-2M stellarator were carried out under the experimental conditions when the ECR heating was the only one heating method used. Time dependences of plasma parameters are given in Figure 1, total heating power of the two gyrotrons being P = 0.4 MW.

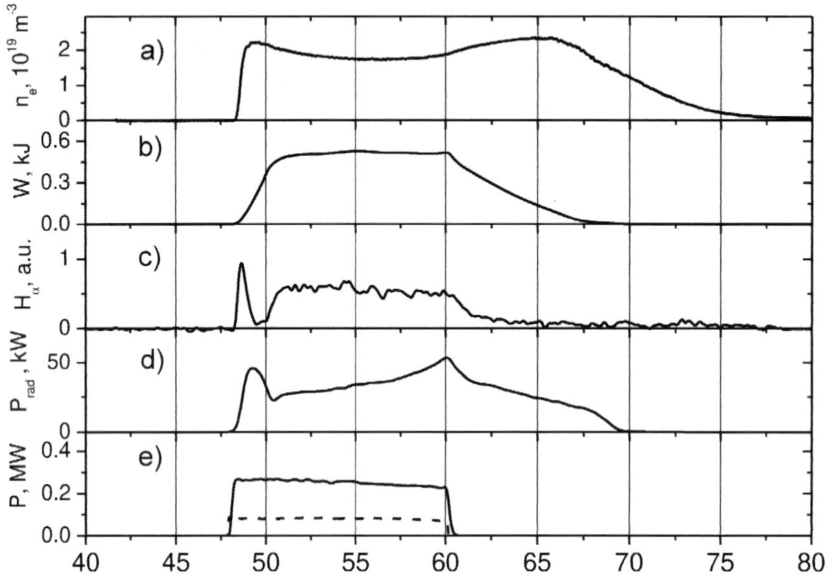

Figure 1. Time development of plasma parameters and ECRH power during the operation pulse #18222: (a) mean electron density; (b) plasma energy contents; (c) H_α line intensity; (d) radiation loss; and (e) input power of the two gyrotrons used for the ECR plasma heating.

At ECRH power of P = 0.1–0.15 MW, the electron density profiles of plasma measured by the multi-chord HCN laser interferometry diagnostics become flattened in the stationary stage of the discharge. They can be approximated by the parabolic function $n_e = n_0{\cdot}(1 - (r/a_p)^p)$, the exponent being $p \approx 8$. Here, ap is plasma radius. Further increase in plasma heating power results in the occurrence of non-monotonous hollow density profiles. Figure 2 presents the density profiles measured at different times under the experimental conditions similar to those shown in Figure 1. As seen from Figure 2, the initial density profile is parabolic (Figure 2, curve 1). In the course of the ECRH pulse, it is gradually transformed into non-monotonous hollow density profile (Figure 2, curve 4). After the end of the ECRH pulse, energy content of plasma gradually decreases during 7–8 ms, the mean density remaining almost unchanged, but the density profiles being transformed (see Figure 1). In two milliseconds after the end of ECRH pulse, the density profile becomes monotonous with the maximum at the axis of the plasma column (Figure 2, curve 5). This is the evidence of the fact that hollow density profiles occur as a result of ECR heating of plasma. At the maximum heating power of P = 0.75 MW, the concavity factor in the central region becomes drastic: $\zeta = 1 - n_e(0)/n_e^{max} = 0.5$.

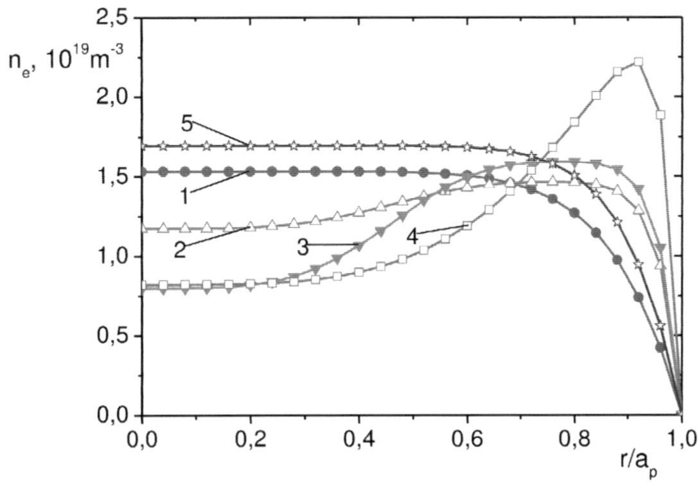

Figure 2. Density profiles in the ECRH experiments measured at different times. (1) t = 50 ms; (2) t = 51ms; (3) t = 52 ms; (4) t = 58 ms; and (5) t = 62 ms (t = 50 and 62ms correspond to the 2nd ms after the ECRH switching on and the 2nd ms after the ECRH switching off, respectively (see Figure 1)).

To our opinion, the electron flow which is directed outward the heating region (the so-called "density pump-out" effect) has a strong effect on the formation of the density profile during a quasi-stationary stage of the heating pulse. The "density pump-out" effect is observed in the ECRH experiments at various tokamaks and stellarators [4, 5]. At the L-2M stellarator, the electron density pump-out effect the most strongly manifests itself due to the high ECRH power density.

3. Flattening of electron temperature profile

With growing ECRH power density, the axial electron temperature measured by the multi-chord soft X-ray (SXR) and the electron cyclotron emission (ECE) diagnostics gradually increases. According to the measurements, at heating powers of P_{ECRH} = 0.25–0.75 MW, radial temperature distributions in the plasma core r/a_p < 0.4 become flat (Figure 3, curve 1).

A flat electron temperature profile may occur as a result of non-axial absorption of ECRH power. One of the possible reasons for non-axial absorption of ECRH power is the deflection of microwave beam from the equatorial plane of plasma. The results of ray-tracing calculations performed for experiments with the axial ECR heating at the L-2M stellarator are shown in Figure 4a. Solid ellipses mark the magnetic surfaces and dotted lines correspond to the lines $|B|$= const. The calculations were performed for the profiles of electron temperature (dashed line) and density (solid lines) shown in Figure 4b. The microwave beam trajectories 1–4 in Figure 4a correspond to the density profiles 1–4 in Figure 4b. Blue segments mark the absorption regions of microwave radiation and red line is the resonance magnetic field line. The calculations prove that, the microwave beams can be considerably deflected from the equatorial plane and some part of microwave radiation can be absorbed not at the plasma axis. In this case, the profiles of the absorbed power and electron temperature can be rather broad.

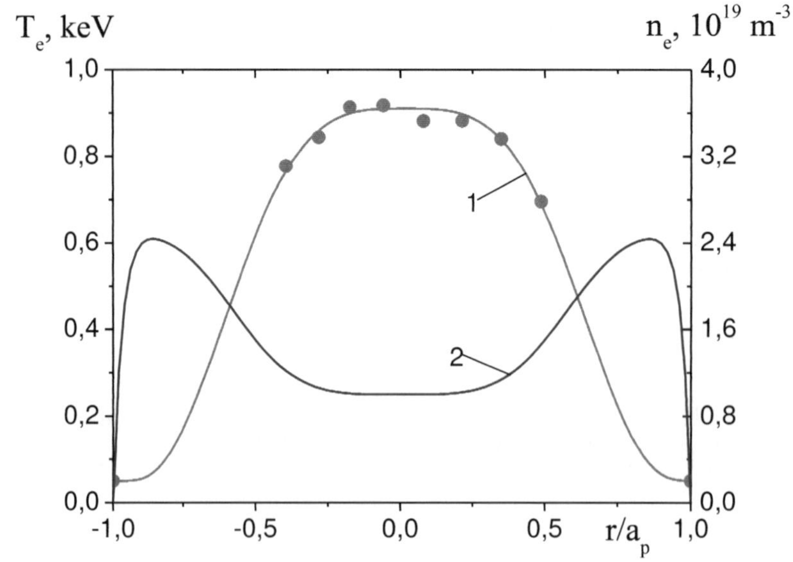

Figure 3. (1) Flattened radial profile of electron temperature and (2) the corresponding electron density profile measured at the ECR heating power of PECRH = 0.45 MW.

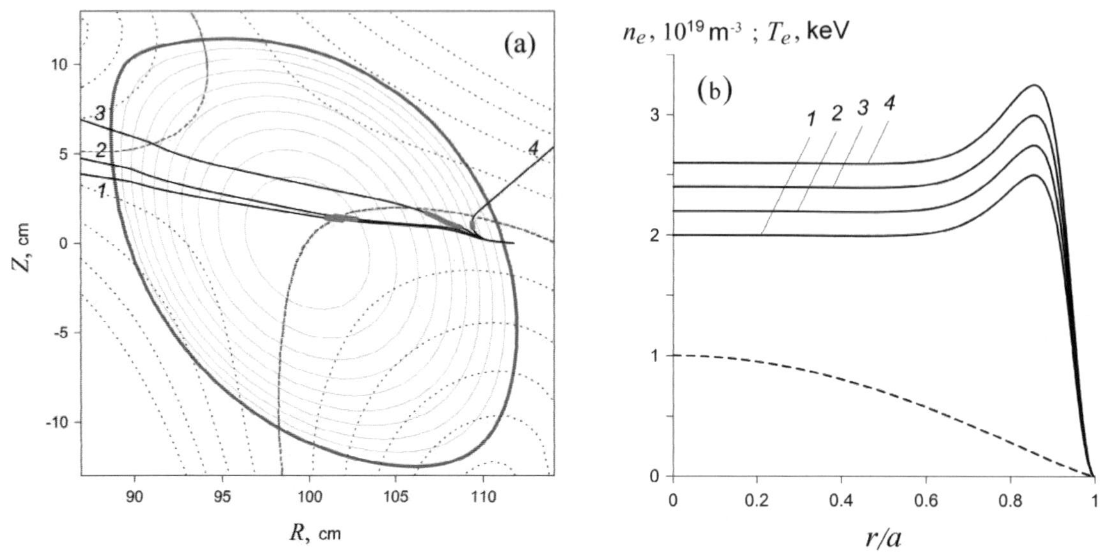

Figure 4. (a) The results of ray-tracing calculations performed for experiments with the axial ECR heating at the L-2M stellarator, (b) the electron temperature (dashed line) and density (solid lines) profiles used in calculations.

4. Observation of non-Maxwellian soft X-ray spectra in experiments on ECR plasma heating

In the experiments on ECR plasma heating with a heating power of P_{ECRH} = (0.1–0.75) MW, the formation of non-Maxwellian electron distribution function over energies was observed.

To measure the distribution functions of electrons over energies, we have used the SXR spectrometer which can measure spectra in the energy range from 1 to 80 keV. A distinctive feature of this spectrometer is its high output count rate $V = 1.7 \cdot 10^5$ counts per second. This characteristic of the spectrometer is very important because, at the L-2M stellarator, the ECRH pulse duration is 12 ms and the duration of the quasi-stationary stage of the operation pulse is about 5 ms. In the energy range in the vicinity of 6 keV, the spectral resolution of this spectrometer was $\Delta E = 320$ eV.

In these experiments, creating and heating of plasma was carried out using one gyrotron with a power of P = 200–600 kW and a frequency of f_1 = 75 GHz. The magnetic field at the axis of the plasma column was B_0 = 1.34 T which corresponds to the location of the resonance region at the plasma axis. Total spectrum was measured during several shots of the facility in the quasi-stationary stage of the discharges (Δt = 55–60 ms) when the basic plasma parameters reach their steady values.

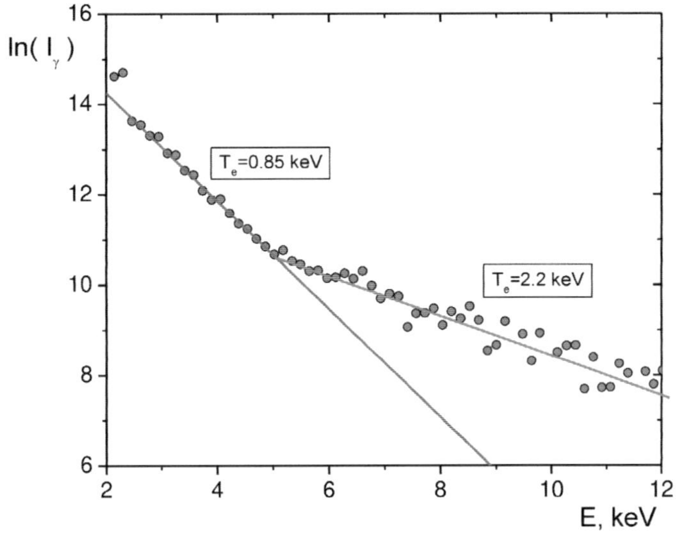

Figure 5. The experimental SXR spectrum reconstructed with allowance for absorption of radiation by a beryllium foil (in a semi-logarithmic scale).

The SXR spectrum reconstructed with allowance for the absorption of radiation by the beryllium foil is shown in Figure 5 in a semi-logarithmic scale. The SXR spectrum was measured along the central chord in the equatorial plane of the stellarator. The mean central-chord plasma density was $n_e = 1.5\cdot10^{19}$ m^{-3} and the heating power was P_{ECRH} = 450 kW. In the energy range of $E/T_e > 7$, this spectrum deviates from the Maxwellian one. It can be approximated by the two straight lines with the slope angles corresponding to the temperatures of the main (thermal) and supra-thermal portions of electrons of T_e^1 = 0.85 keV and T_e^2 = 2.2 keV, respectively. The salient point of the spectrum corresponds to the energy of E = 5 keV.

The similar two-slope SXR spectra were measured in experiments on the ECR plasma heating and current drive with high specific heating power at several tokamaks, e.g. at the T-10 [6] and TCV [7] tokamaks. Generally, in these experiments, the specific heating power heating power was high. The authors of [6, 7] suggested no explanation for the formation of two-slope spectra.

Experiments intended for determination of the plasma regions which are sources of supra-thermal electron emission were carried out at the L-2M stellarator. It was expected that the plasma heating region (region of absorption of microwave radiation) is the main source of supra-thermal electrons. A scanning system was used to measure SXR spectra along different chords in the vertical section of the plasma column in the range of impact parameters of $-0.45 < r/a < 0.65$ (see Figure 6). The straight line in Figure 6 corresponds to the Maxwellian spectrum with a temperature of T_e = 550 eV which corresponds to the thermal portion of the spectrum measured at a radius of $r/a = -0.45$. Negative radii correspond to the chords in the bottom half-plane of the plasma cross section. All the spectra measured along the chords in the bottom half-plane turned out to be almost Maxwellian. In the upper half-plane, a supra-thermal "tail" appears in the spectra. Its temperature and intensity increase with growing distance from the central chord and reach their maximum values at a radius of r/a = 0.55 cm. When the measurements are carried out along this chord, the region in the vicinity of the separatrix X-point falls into the field of view of the spectrometer, where the plasma-wall interaction is the most intense. At longer distances from the plasma axis, the deviations of the measured spectra from the Maxwellian one become slightly smaller. The measurements performed show that, in the experiments on the axial ECR plasma heating, the supra-thermal tail of SXR spectrum is not mainly formed by electrons radiating in the plasma heating region. This issue still remains the subject of further research.

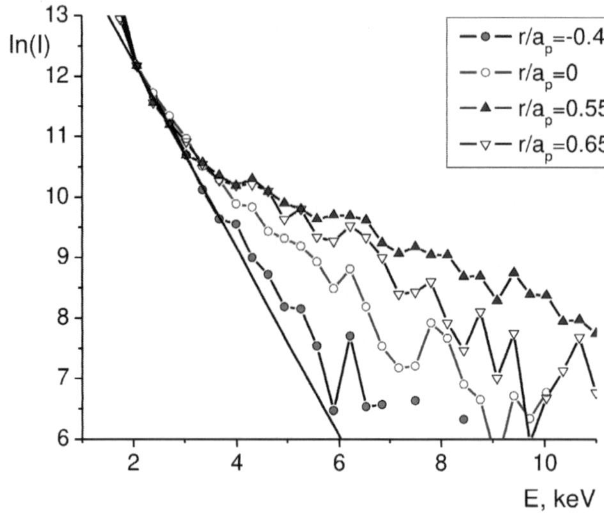

Figure 6. SXR spectra measured along chords with different impact parameters r/a.

5. Plasma confinement and energy lifetime data

In experiments with the maximum power density of $P/V = 3.0$ MW/m^3, plasma density and radiation loss temporally increase during the operation shot. As the gas puffing into the L-2M chamber is switched off 30 ms before the plasma formation, this increase is a result of intensification of the plasma-wall interaction. But this fact does not negatively affect the plasma confinement. Figure 7 shows the comparison between the energy lifetime data measured in experiments with ECRH power of up to $P_{ECRH} = 0.75$ MW at the L-2M stellarator and the energy lifetime data calculated according to the LHD scaling. White and black points correspond to the experimental data obtained in the shots with microwave power up to $P_{ECRH} = 0.4$ MW and 0.4 MW $< P_{ECRH} < 0.75$ MW, respectively. The red line corresponds to the scaling obtained previously basing on the experimental data of the L-2M stellarator [8]:

$$\tau_E^{L\text{-}2M}[\text{s}] = A \cdot n_e^{0.71}[10^{20}\ \text{m}^{-3}] \cdot P^{-0.69}[\text{MW}],$$

where A = 2.32·10-3. It slightly differs from the LHD scaling (black line in Figure 7) [9]:

$$\tau_E^{LHD}[\text{s}] = 0.17\ B^{0.84}[\text{T}]\ a^{2.0}[\text{m}]\ R^{0.75}[\text{m}]\ n_e^{0.69}[10^{20}\ \text{m}^{-3}] \cdot P^{-0.58}[\text{MW}],$$

particularly, the dependences on the heating power are different. The energy lifetimes measured at the L-2M stellarator are consistent with those determined from the LHD scaling within 15%.

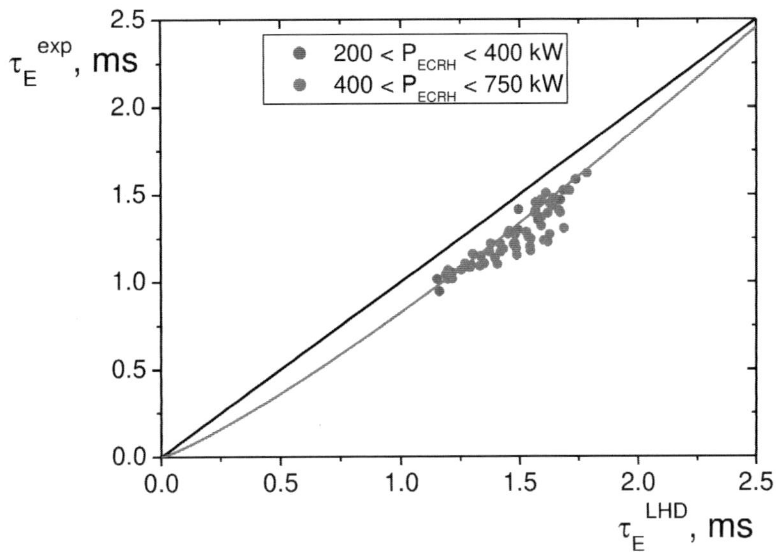

Figure 7. Energy lifetime data measured in experiments carried out with various plasma heating powers at the L-2M stellarator.

Despite the increasing plasma-wall interaction, we have not observed the dramatic deterioration of plasma confinement at the L-2M stellarator when the ECRH power grows up to P_{ECRH} = 0.75 MW (ECRH power density increases up to 3.0 MW/m^3). As seen in Figure 8, there is no difference in experimental energy lifetimes in shots with low (P_{ECRH} < 0.4 MW) and high (0.4 MW < P_{ECRH} < 0.75 MW) heating powers.

6. Conclusions

The experiments on ECR plasma heating were carried out at the L-2M stellarator at high volume power density (up to 3.0 MW/m^3). Under these conditions, non-monotonous hollow density profiles were measured. At a maximum heating power density of 3.0 MW/m^3, the concavity in the axial region becomes drastic $n_e(0)/n_e^{max}$ = 0.5. The formation of hollow density profiles can be associated with "density-pump-out" effect which the most strongly manifests itself at the L-2M stellarator due to high ECRH power density.

In these experiments, plasma temperature profiles measured in the core plasma region r/a_p < 0.4 occur to be rather flat. A flat electron temperature profile may occur as a result of non-axial absorption of ECRH power.

Two-slope SXR spectra were also measured in the energy range from 2 to 12 keV. In the energy range of E/T_e > 7, the formation of supra-thermal tails was observed in the spectra. Experiments intended for determination of the plasma regions which are sources of supra-thermal electron emission were carried out. It was revealed that the supra-thermal tails of SXR spectra are not mainly formed by electrons radiating in the plasma heating region.

Studies of plasma confinement at high ECRH power densities were carried out. With ECRH volume power density growing up to 3.0 MW/m^3, we have not observed dramatic deterioration of plasma confinement at the L-2M stellarator. The measured energy lifetime is generally consistent with that determined from the LHD scaling.

Further we plan to continue studies on plasma confinement in experiments on ECR plasma heating using two gyrotrons with total microwave power up to P_{ECRH} = 1.0 MW and study the plasma confinement at still higher power densities.

Acknowledgments
This work was supported in part by the Russian Federal Agency for Scientific Organizations (state assignment no. 01200953485) and Russian Academy of Sciences under the Program I.11П.

References

[1] Shchepetov S V, Kholnov Yu V, Fedyanin O I, Kuznetsov A B, Vasilkov D G, Akulina D K, Batanov G M, Gladkov G A, Grebenshchikov S E and Meshcheryakov A I Sandwich structure of plasma edge during transition to improved confinement regime in L-2M stellarator 2008 *Plasma Phys. Control. Fusion* **50** 045001 (16pp) doi:10.1088/0741-3335/50/4/045001
[2] Batanov G M, Belousov V I, Bondar Yu F et al, A New MIG-3 Gyrotron Complex for Creation and Heating of Plasma in the L-2M Stellarator and the First Experimental Results 2013 *Plasma Physics Reports* **39** 13 1088
[3] Meshcheryakov A I, Akulina D K, Batanov G M et al, Effect of vacuum chamber boronization on the plasma parameters in the L-2M stellarator 2005 *Plasma Physics Reports* **31** 6 452
[4] Erckmann V and Gasparino U 1994 *Plasma Phys. Control. Fusion* **36** 1869
[5] Ryohei Makino, Shin Kubo, Takeshi Ido et al Local and Fast Density Pump-out by ECRH in the LHD 2013 *Plasma and Fusion Research: Regular Articles* **8** 2402115
[6] Esipchuk Yu V, Kirneva N A, Martynov A A and Trukhin V M Suprathermal SXR emission in ECCD experiments at the T-10 tokamak 1995 *Plasma Physics Reports* **21**7 577–583
[7] Nikkola P, Sauter O, Behn R, Coda S, Condrea I., Goodman T P, Henderson M A, Hervey R W and TCV team 2003 *Nucl. Fussion* **43** 1343-1352
[8] Fedianin O I, Akulina D K, Batanov G M et al 2007 *Plasma Physics Reports* **33** 10 1
[9] Sudo S, Takeiri Y, Zushi H, Sano F, Itoh K, Kondo K and Hyoshi A. 1990 *Nuclear Fusion* **30** 1 11–21

IOP Conf. Series: Journal of Physics: Conf. Series **907** (2017) 012017 doi:10.1088/1742-6596/907/1/012017

Features of design of the Doppler reflectometry diagnostics at the L-2M stellarator intended for operation under conditions of high ECRH power

A A Kharchevskii[1,2], N N Bogachev[1,2,3,6], D V Malakhov[1,2,3,5], V I Nefedov[2] and N N Skvortsova[1,4]

[1] Prokhorov General Physics Institute, Russian Academy of Science, 38 Vavilov street, Moscow, 119991, Russia
[2] Moscow Technological University, 78 Vernandsky prosp., Moscow, 119454, Russia
[3] Pirogov Russian National Research Medical University, Ostrovitianov str. 1, Moscow, 117997, Russia
[4] National Research Nuclear University MEPhI, Kashirskoe shosse, 31, Moscow, 115409, Russia

[5] 89168766306@mail.ru, [6] bgniknik@yandex.ru

Abstract. Features of design of the Doppler reflectometry diagnostics at the L-2M stellarator are discussed. This contact-free diagnostics is intended for experiments carried out under conditions of high-power microwave plasma heating at a frequency of electron cyclotron resonance. Resources of the proposed Doppler reflectometry scheme were studied with allowance for the installation of a new system of rejection microwave filters involving the Fabry–Perot resonance filter and pin-shaped waveguide filters. The simulation results of the resonator-filter operation as well as the results of test measurements using the MVNA-8-350 millimeter vector network analyzer are presented.

1. Introduction

Contact-free diagnostic methods are very important for the fusion facilities with the magnetic confinement of high-temperature plasma. The Doppler reflectometry (DR) diagnostics [1] is such a contact-free technique that provides information on poloidal rotation of plasma basing on the frequency shift of the backscattered radiation spectrum; it is used at many tokamaks and stellarators.

One of the overall used methods of plasma heating is microwave plasma heating at a frequency of electron cyclotron resonance (ECR). A gyrotron complex with a power of up to 24 MW at a frequency of 170 GHz will be used at the International Tokamak-Reactor (ITER) for additional microwave plasma heating. The MIG-3 gyrotron complex with a power of up to 1 MW for ECR plasma heating at a frequency of 75.3 GHz (the second harmonic of gyrotron frequency) was put into operation at the L-2M stellarator [2]. The launch of the MIG-3 complex made it possible to achieve a record to date specific power of 4 MW/m^3 which can be inputted into plasma, which strongly exceeded those achieved at the other toroidal fusion facilities: up to 0.4 and 0.04 MW/m^3 at the T-10 tokamak and the LHD stellarator, respectively. In the experiments at the L-2M stellarator, it was found that, at such high ECRH specific power, the spurious signal at a working frequency of the gyrotron complex exceeds the useful signal in the receiving path of the DR diagnostics [3].

The goal of the research was to provide the stable operation of the DR diagnostics in experiments on high-power ECR plasma heating. To reduce the effect of an external signal at a frequency of 75.3 GHz and noise signal in a frequency range of range 60–80 GHz, it was necessary to create a new system of filters. In the new system of filters, we have used band-stop filters. These filters can provide strong signal attenuation (suppression) in a certain frequency range Δf near the central frequency f_0 [4].

Because of the design features of the diagnostics (in particular, the small overall dimensions of the waveguide paths), the installation of additional pin-shaped filters, similar to those in [1], was not possible. It was required to design and manufacture a compact band-stop filter with a central suppression frequency of 75 GHz. We have designed a filter consisting of the Fabry–Perot resonators arranged succesively. In the millimeter and submillimeter frequency range, the Fabry–Perot resonance filters are actively used in various diagnostics and radioastronomical instruments [5–8]. In [5, 6], two-dimensional flat metal grids deposited on transparent insulating bases are used as reflectors. The characteristics of such filters strongly depend on the accuracy and quality of the grid deposition. We used the Fabry–Perot resonators with flat round mica plates (dielectric constant of mica is $\varepsilon = 6$) with a diameter of D = 80 mm and a thickness of S = 0.1 mm.

Along with the modification of the DR diagnostics, the optimization of data analysis algorithms was performed and the new software was developed. The first test measurements showed an improvement of the signal-to-noise ratio, which made it possible to use the DR diagnostics in the experiments with ECRH powers of up to 1 MW at the L-2M stellarator.

1.1. Pin-shaped waveguide filters

Three band-stop filters (F2, F3, F4 in Figure 1) based on half-wave resonators were made using 8-mm-wide waveguides. Metal pins with a diameter of 0.9 mm were inserted into the waveguides perpendicularly to their long side. The number of pins required was determined experimentally to achieve the necessary suppression ratio. As a result, the F2 filter consists of 40 pins, and each of the F3 and F4 filters consist of 30 pins. The distance between the pins (waveguide inhomogeneities) was calculated as half-wavelength of the gyrotron wave in the waveguide, which is a noise disturbance for the diagnostics:

$$\lambda_{\rm H} = \frac{\lambda}{\sqrt{\varepsilon_\mu + \left(\frac{\lambda}{\lambda_{cr}}\right)^2}} \tag{1}$$

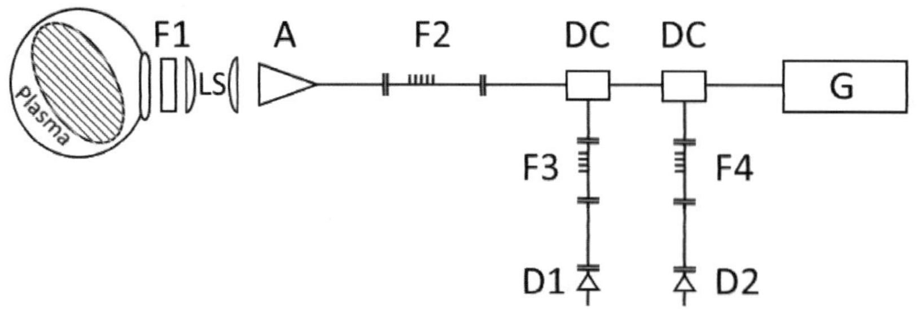

Figure 1. Schematics of the Doppler reflectometer filtering system. (A) antenna; (LS) lens system; (F1) the Fabry–Perot resonance filter; (F2), (F3), and (F4) pin-shaped waveguide filters; (DC) directional coupler; (D1), (D2) detector heads, and (G) microwave generator.

Here, λ_H is wavelength of the gyrotron extraneous wave in the waveguide; ε_μ is the dielectric constant, we take it equal to 1, since the waveguide is filled with air; λ is the wavelength of the extraneous wave with the interfering frequency in vacuum; λ_{cr} is the critical wavelength of the wave

IOP Conf. Series: Journal of Physics: Conf. Series **907** (2017) 012017 doi:10.1088/1742-6596/907/1/012017

that can propagate in a 8-mm-wide waveguide. Taking into account that, in our experiments, the H_{10} wave is generated, we obtain the critical wavelength of λ_{cr} = 14.4 mm. Diameter of pins is a = 0.9 mm, and so the distance between the pins is 2.07 mm.

1.2. Fabry–Perot Resonance Filter

The proposed filter is the Fabry–Perot resonator-interferometer with two parallel mica plate reflectors. To obtain a band-stop filter, we need to have an interference minimum at a frequency of 75 GHz (λ_0 = 4 mm). We can use the well-known relation:

$$2d_0 = m\lambda_0, \tag{2}$$

where d_0 is the resonator length in vacuum, m is integer number, λ_0 is length of an electromagnetic wave in vacuum. So, the resonator length is:

$$d_0 = m\lambda_0/2, \tag{3}$$

For m=1 and λ_0 = 4 mm, the resonator length is d_0 = 2 mm. However, in our case, we use mica plates with a permittivity of ε = 6 and a thickness of S = 0.1 mm. These parameters should be taken into account when calculating the resonator length. The wavelength of an electromagnetic wave in mica is equal to:

$$\lambda = \frac{\lambda_0}{\sqrt{\varepsilon}}. \tag{4}$$

With allowance for relation (4), the resonator length d can be calculated as:

$$d = d_0 - \Delta d, \tag{5}$$

where Δd is the difference between the resonator length in vacuum and the length of the resonator with allowance for mica plates. It can be calculated as follows:

$$\Delta d = m\frac{nS(\lambda_0 - \lambda)}{2}, \tag{6}$$

where m is an integer (m = 1, in our case), n is the number of mica plates in one resonator (n = 1), and S is thickness of mica plates (S = 0.1 mm). Thus, we get $\Delta d \approx 0.3$ mm, d = 1.7 mm and the distance between plates in the resonator is l = 1.5 mm. The filter with four Fabry–Perot resonators was chosen to be uses in the filtration system.

1.3. Numerical modelling of the Fabry–Perot resonance filter.

The 3D-model (XYZ coordinate system) of the Fabry–Perot resonance filter was created using the computer-aided design system Electro-Magnetic Professional (CAD EMPro) from Keysight Technologies (formerly Agilent Technologies). The band-stop filter with the Fabri-Perot resonators was simulated by several pairs of coupled mica plates, arranged one after another at a distance of l from each other. The plates were installed inside a coaxial waveguide perpendicularly to its axis to ensure the propagation of TEM wave. The calculations were carried out using the Finite Element Method (FEM) by the Agilent FEM Simulator block [9], the boundary conditions at the edges of the countable domain were set in the form of a Perfect Matching Layer (PML). Waveguide ports were installed on both sides of the filter.

2. Simulation and measurements results

The filter characteristics were measured with a frequency step of 10 kHz using the MVNA-8-350 vector network analyzer from AB Millimetre Company.

The losses (or transmission coefficient S21) of the pin-shaped waveguide filters were measured in the ranges of 29–41 GHz (operating frequency range of the DR diagnostics) and 62.5–100 GHz. The transmission coefficient of the pin-shaped filter does not exceed 0.5 dB in the operating frequency range of the Doppler reflectometer and it is more than 30 dB at the ECR frequency (75.3 GHz). Taking into account the fact that three waveguide filters are used in the filter system (F2, F3 and F4 in Figure 1), the suppression of the ECRH frequency (75.3 GHz) in the waveguide section of the filter system should be about 60 dB in each detector channel.

Using the model, the transmission coefficient of the Fabry–Perot filter was calculated in the frequency range of 40–80 GHz. Figure 2 shows the results for the optimized model (curve 2). As can be seen, at a frequency of 75.3 GHz, the attenuation coefficient S21 = –21.9 dB was obtained. We note that the attenuation at frequencies of 35 and 40 GHz calculated using the optimized filter model does not exceed –0.9 dB. Figure 2 also shows the measured transmission coefficients of the Fabry–Perot resonance filter. Due to the specifics of the used analyzer, its readings can be taken only at discrete spectral ranges. It should be taken into account that all measurements were carried out under conditions of the continuous generation of low-power radiation in MVNA-8-350 vector network analyzer (about some hundreds of milliwatts).

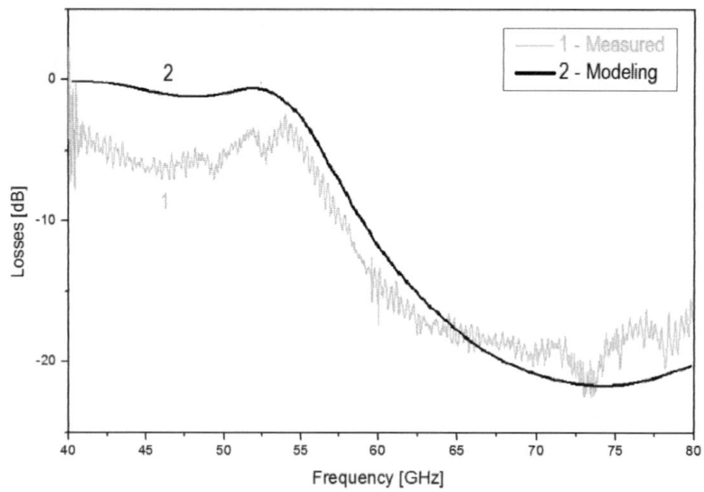

Figure 2. The losses in the Fabry–Perot resonance filter. Curves 1 and 2 correspond to the measured and calculated filter attenuation, respectively.

3. Conclusions

The first experiments at the L-2M stellarator using the upgraded Doppler reflectometer showed that, due to optimization of the filter system, the gyrotron radiation is suppressed in the detector channels at least by 80 dB. The presented diagnostics scheme provides conducting experiments on ECR plasma heating using gyrotrons with a power of up to 1 MW.

In the course of optimization of the Doppler reflectometry diagnostics at the L-2M stellarator, the band-stop Fabry–Perot resonance filter for the frequency band of 70–80 GHz made of doubled mica plates was simulated for the first time. The measured suppression coefficient of the filter in a given frequency band is in good agreement with calculated one. We received evidence that when the wavelength of the resonator changes even by a few tenths of millimeter, the filter transmission band can also change, so we note that it is very important to observe the accuracy of resonator dimensions when manufacturing the resonator. Measurements of the filter characteristics showed that the actual suppression coefficient S21 of the manufactured filter is −19.5 dB at a frequency of 75 GHz which is close to the calculated one (−21.9 dB).

IOP Conf. Series: Journal of Physics: Conf. Series **907** (2017) 012017 doi:10.1088/1742-6596/907/1/012017

Acknowledgments
The authors are grateful to D.A. Shelukhin for help in the organization of measurements. The work was supported in part by Ministry of Science and Education of Russian Federation (RF President grant for state support of young Russian scientists no. MK-5298.2016.8).

References

[1] Batanov G M, Belousov V I, Bondar Yu F, et al 2013 *Plasma Phys. Rep.* **39**(13) 1088
[2] Pshenichnikov A A, Kolik LV, Malykh N I, et al 2005 *Plasma Phys. Rep.* **31**(7) 554
[3] Kharchevskii A A, Bogachev N N, and Malakhov D V 2016 *Engineering Physics (Inzhenernaya Fizika)* **6** 59 (in Russian)
[4] Wannier P G, Arnaud J A, Pelow F A and Saleh A.A 1976 *Rev. of Scientific Instr.* **47**(1) 56
[5] Arzhannikov A V, Ginzburg N S, et al 2006 *Vestnik of NSU: Physics Series* **1**(2) 71
[6] Clarke R.N. and Rosenberg C.B. 1982 *Journal of Physics* E: *Scientific Instruments* **15** 9
[7] Vinogradov E.A., Dianov E.M., and Irisova N.A. 1965 *Sov. J. of Exp. Theor. Phys. Lett.* **2** 205
[8] Hunter I 2001Theory *and design of microwave filters IEE: Electromagnetic wave series* **48**
[9] Keysight Technologies. About EMPro url: http://www.keysight.com/en/pc-1297143/EMPro

IOP Conf. Series: Journal of Physics: Conf. Series **907** (2017) 012018 doi:10.1088/1742-6596/907/1/012018

Metal nano-particles modernized layers and those with polymers for laser thermonuclear targets

I V Akimova[1], A A Akunets[1], N G Borisenko[1], S Chaurasia[2], A I Gromov[1,3], C Kaur[2], D S Munda[2], A S Orekhov[1], G V Sklizkov[1], S M Tolokonnikov[1], U Rao[2] and V Rastogi[2]

[1] P N Lebedev Physical Institute (LPI), Leninskiy pr. 53, Moscow 119991, Russia
[2] Bhabha Atomic Research Center (BARC), Trombay, Mumbai-85, India

[3] E-mail: agrom@sci.lebedev.ru

Abstract. The manufacturing and precision monitoring methods of the layers as promising direct and indirect targets for Inertial confinement fusion (ICF) are under study, as well as their application in the experiments. The metal-containing foams with a density that is several times or several orders of magnitude smaller than the full-density material of the same composition are of interest for higher laser light conversion into X-rays and for better energy delivery into the target in direct and indirect interaction schemes. Such targets are developed and provided. We report the interaction of Nd: glass laser with a low-density bismuth and gold targets. The plasma dynamics and X-ray emissions were observed using multiframe optical shadowgraphy and an X-ray streak camera. Enhanced X-ray intensities and festoon plasma flame are observed from the metal low-density layers.

1. Introduction

The analysis of laser radiation conversion has been presented in the publication by S. Yu. Gus'kov and Yu. A. Merkul'ev [1], where the authors considered the problems of conversion efficiency in a low-density polymer foam with an addition of nanoparticles of heavy elements, and the absorption that takes place in the external layer of the target. This layer plays the role of the laser radiation absorber and the converter of laser radiation into X-rays. The degree of conversion increases with the decrease of the laser wavelength, and for the intensity of 10^{14}-10^{15} W/cm^2, the degree of laser energy conversion is 60-80% [2].

In [3], the authors have theoretically predicted the increase of temperature in the converter due to the X-ray radiation when a foam with a low-density gold layer is introduced instead of solid gold. As shown in [4], the experimental verification of this important statement using a laser facility has confirmed the validity of this statement.

In addition, such layers may be helpful for the formation of spherical layers for the cryogenic targets and low-density media in the form of metal-polymer foams and ultra-dispersive powders (UDP) of metal, which are used both as free layers and as additives in the foam polymers.

The experimental studies at BARC, Mumbai, India, in collaboration with scientists from the Thermonuclear target laboratory, LPI, Moscow, at the Nd: Glass laser facility (wavelength, 1.062 μm; energy, 15 J; intensity, ~10^{14} W/cm^2; duration, 0.5 ns) provided important results.

IOP Conf. Series: Journal of Physics: Conf. Series **907** (2017) 012018 doi:10.1088/1742-6596/907/1/012018

2. Experiment

In our previous experiment, using a low-density layer of ultra-dispersive (UDP) bismuth powder (0.1 g/cm^3), the conversion efficiency increased by 10% compared with the solid matter target in the soft X-ray region [5]. The physics of conversion of laser radiation into X-rays is discussed in our paper [6]. In addition, the abovementioned paper deals with a number of important technological aspects.

The Organic Chemistry Institute and LPI of the Russian Academy of Sciences has developed polymer aerogel layers with heavy metal nanoparticles. A multistep technological system of adding a metal cluster to the polymer is used [7]. Currently, similar studies are being continued and carried out based on the novel technological developments [8-9]. It is also noted that the parylene high-strength plastic (C_8H_8) is used, as a rule, in the form of a micrometer-thick protective coating of micro-circuit inner surfaces. The plastic may contain Br-parylene (C_8H_7Br) in a form that can be used for diagnostics in addition to the reaction of ^{79}Br(n,2n)^{78}Br, which takes place in laser thermonuclear fusion targets with a reaction cross-section of 862 mbarn, $t_{1/2} = 6.5$ min, E γ-quanta 0.51 MeV. We have succeeded in obtaining a low-density porous material with densities of (1/3-1/12)th of solid C_8H_7Br. This is of interest for the diagnostics in thermonuclear experiments (Fig. 1a).

The development of coating deposition technology has allowed us to obtain more uniform and ultradispersive Bi and Au layers. The scanning electron microscope (SEM) image of UDP Bi with ~40 nm particles is shown in Fig. 1b.

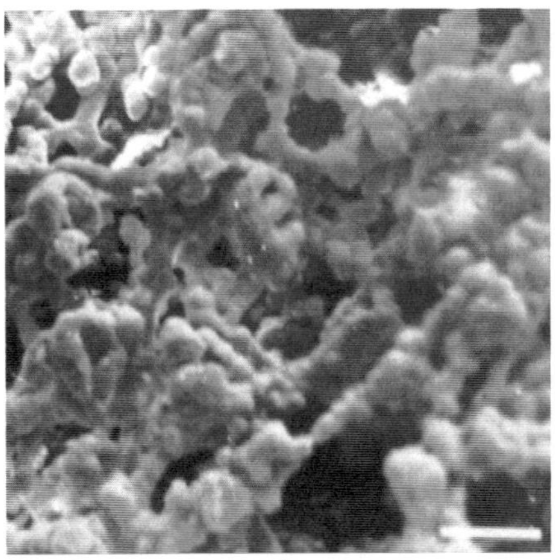

a)

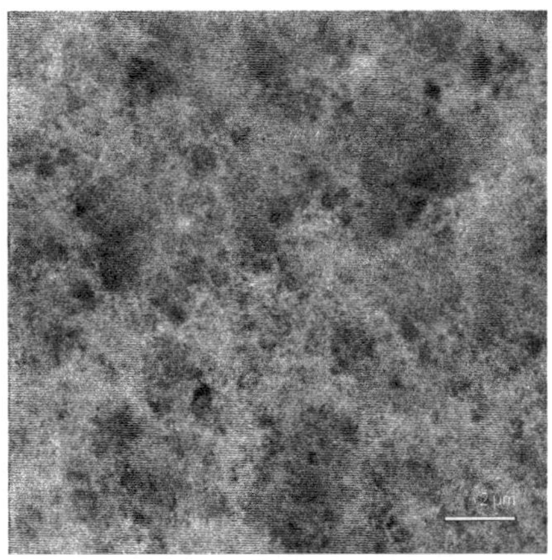

b)

Figure 1. SEM image of (a) low-density parylene foams that are doped with bromine (scale, 10 μm) and (b) ultradispersive ~40 nm Bi particles (scale, 2μm).

3. Results and discussion

The size of nanoparticles and density of layers depend on the pressure change of inert gas (Ar or He) that is used in the manufacturing installation. Based on our experience, in the case of He, we obtain a finer ultradispersive powder, smaller density of the layer, temperature rise regime, as well as a shorter time of the evaporation process. Besides, there is a dependence on the amount of evaporated matter and the type of evaporator, which is our exclusive development. Recently, we managed to deposit a uniform layer of Bi UDP (a 100-μm thickness and a density as low as 1/250th of full-density Bi) on a 0.5-μm thick cellulose nitrate film, which was placed on a holder with a 32-mm inner diameter opening, as shown in Fig 2a. The gold ultradispersive layer with a particle size of 30 nm is also obtained using an improved technique that is aimed at producing a finer powder, as shown in Fig. 2b.

Laser investigations were performed using the laser setup shown in Fig. 3.

IOP Conf. Series: Journal of Physics: Conf. Series **907** (2017) 012018 doi:10.1088/1742-6596/907/1/012018

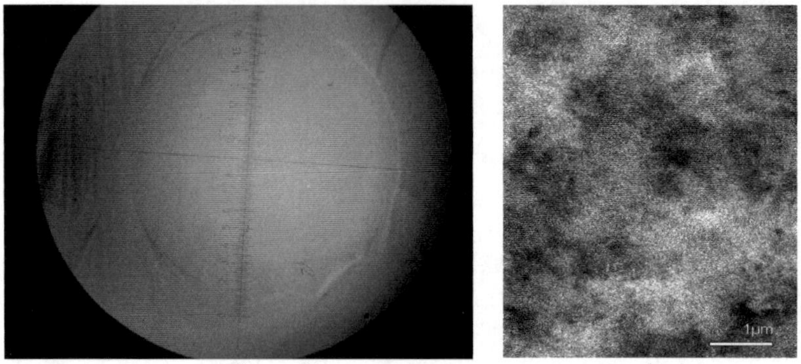

a) b)

Figure 2. (a) The 100-μm bilayer of UDP on a 0.5-μm cellulose nitrate foil, free and uniform over a 32-mm opening in the holder. The scale unit is 2 mm (between two ciphers). (b) SEM picture of a Au-modified UDP layer. The fine particle size is ~ 30 nm. Scale, 1 μm.

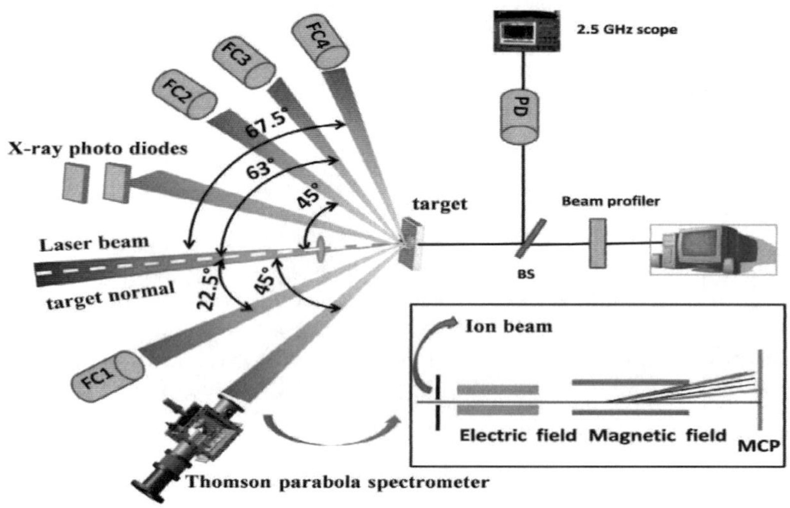

Figure 3. Schematic of the experimental setup for the foam target irradiation with high-power laser pulses.

Shadow photographs supplement the study into the stability of foam target and full-density target plasma (Fig. 4). The vertical strip is a shadow photo of the target profile under the action of laser radiation from the left. The shadow enlarges due to the formation of dense plasma and target-evaporated matter. The normal incidence laser beam initiates characteristic and repeatable, almost periodic, festoon picture and helps study radiation absorption and conversion together with hydrostability trends of low-density layers application in ICF. Green beam is used for the plasma shadowgraphy.

To interpret correctly the results from the ICF experiments, it is necessary to know exactly the constructional and physical parameters of the targets [10]. An important problem is a precision control of the metal ultradispersive layers, and among them, those in the combined layers of low-density metal polymers to be used in the ICF. The optical and X-ray methods of the layer characterization are used, including microradiography and X-ray tomography, scanning electron microscopy, and radiography. The monitoring methods are mutually complementary.

For comparison, a low-density CD_2 layer (300 μm thickness and the density of 1/3 of the solid polymer), which is mounted on the same holder as above and under the same irradiation conditions, was irradiated. The polymer targets show even and hydrodynamically tranquil flow (Fig. 5), and the festoon features are absent.

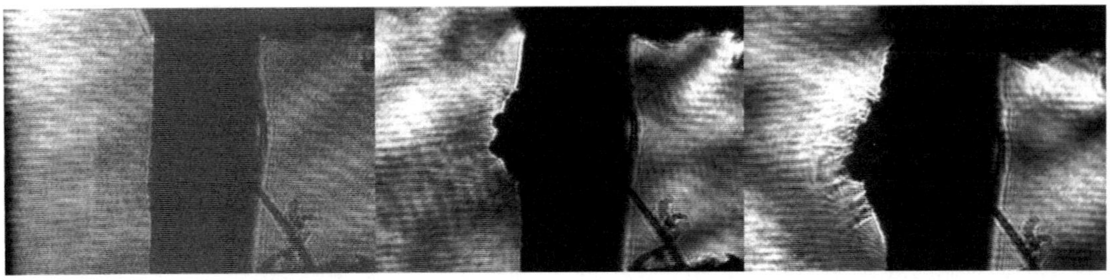

a) b) c)

Figure 4. Plasma expansion from low-density Bi (thickness, 70 μm; density, 1/100 of solid Bi) on a 5-μm-thick Teflon substrate. (a): before the laser pulse; (b): 2 ns after the laser pulse start; (c): 8 ns after the laser pulse.

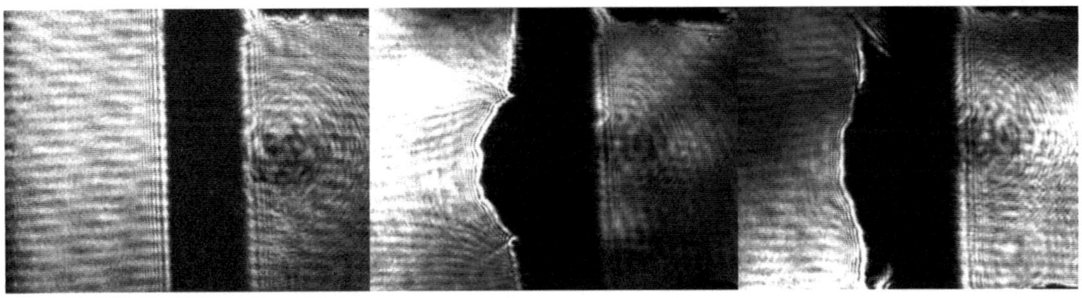

a) b) c)

Figure 5. Plasma expansion from a low-density CD_2 target. (a) before the laser pulse; (b) 2 ns after the laser pulse is initiated; (c) 8 ns after the laser pulse.

This allows one to make quantitative comparison and calculate longitudinal and transverse motion of the waves and matter. Metal foams behave differently (Fig. 4), and it is typical that there are almost periodical perturbations of the opacity front along with the integrally united transverse and axial motion. The motion behavior does not strongly depend on the metal, but is notably dependent on the target matter structure: solid matter, massive or thin foil.

The X-ray emittance of Au foams was compared with that of bulk gold. In Fig. 6, color represents the intensity change in the signals from the foam with a $1/100^{th}$ the density of bulk gold. The X-ray intensity curves are given above for each streak-image. The target matter has a noticeable spatial effect. At the maximum, the X-ray intensity is approximately 15% higher for the low-density metal layer than for the full-density one. Further data processing is required. This will also ensure the foam application for intensive particle generation [11] for the sake of science and ICF.

Currently, there is no obvious radiating capacity drop for the gold foam due to the presence of admixtures with a low atomic number. The possible explanation may be in the current fabrication technique, which ensures that the foam composition is close to that of the initial pure metal.

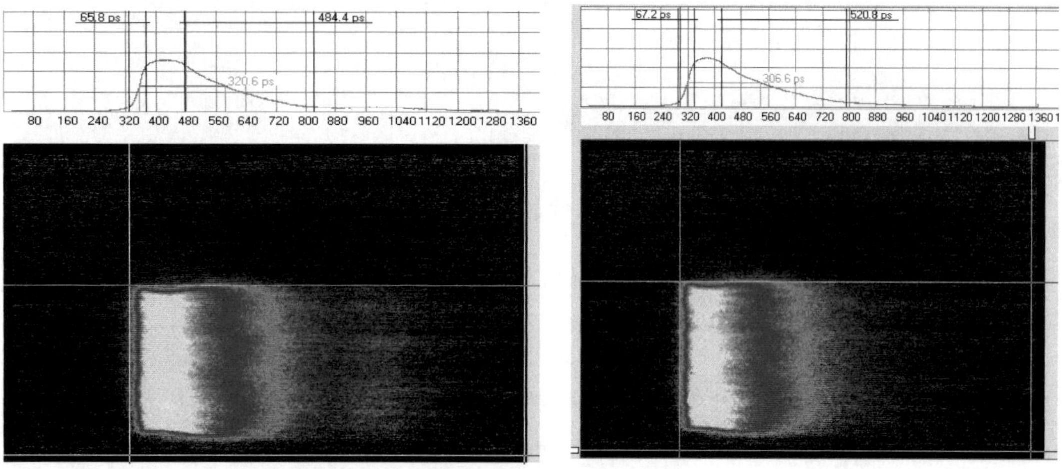

a) b)

Figure 6. Signal from an X-ray streak camera for Au targets with a density of 200 mg/cc (a) and for Au targets with a density of 19.3 g/cc (b). Densitograms for both are placed above each signal.

4. Conclusions

The techniques aimed to obtain the low-density metal layers and metal-polymer foams for the purposes of ICF in a wide density range (5-500 mg/cm^3) have been developed. The discussed layers can be used to obtain a more effective conversion of laser radiation into X-rays, and the layers are stable enough for the transportation and mounting into the chamber at the laser facility with a subsequent vacuum pumping.

Materials that are better than the earlier UDP low-density layers of gold and bismuth on freestanding submicron foils have been developed, manufactured as plain targets, characterized and provided for laser exposure.

The experiments with metal targets, as compared with the low-density metals of the same composition, have confirmed the expected increase of X-ray radiation from the laser plasma of the initial foam target. The observed stable and repeatable hydrodynamic behavior demonstrated specific festoon features as compared with similar polymer layers. The abovementioned characteristics are promising for ongoing and future experiments.

Acknowledgements
The authors are grateful to the Russian Foundation for Basic Research for the partial support of this work: Grants nos. 15-52-45116 and 17-02-00366/17.

References

[1] Gus'kov S Yu and Merkul'ev Yu A 2001 *Quantum Electronics* **31** №4 311-317
[2] Chaurasia *et al* 2015 *Optics Communication* **343** 1-5
[3] Rosen M D and Hammer J H 2005 *Phys. Rev.* E **72** 056403 1-5
[4] Young P E *et al* 2008 *Phys. Rev. Let.* **101** 035001 1-4
[5] Borisenko N G *et al* 2011 *Preprint FIAN* **29**
[6] Akimova I V *et al* 2012 *Problems of atomic science and technology. Series thermonuclear fusion* Issue **2** 122-130
[7] Borisenko N G *et al* 2006 *J de Physique* IV (France) **133** 305-308
[8] Borisenko N G *et al* 2013 Laser study into and explanation of the direct-indirect target concept *EPJ Web of Conferences* **59** 03014 1-5
[9] Orekhov A S *et al* 2016 Modern trends in low-density materials for fusion *J of Physics:*

Conference Series **688** (1) 012080 (after IFSA 2013, Nara, Japan, *Book of Abstr* p 315)

[10] Borisenko L A *et al* 2014. Metal produced as nano-snow layers for converters of laser light for indirect targets as intensive EUVsourses *J of Radioanalitical and Nuclear Chemistry* **299** №2 955-960

[11] Borisenko L A *et al* 2016 Distribution function evolution of electrons stochastically heated by picosecond laser pulse ECLIM 34 2016 Moscow Russia *Book of Abstr.* 18 http://plasma.mephi.ru/ru/uploads/files/conferences/ECLIM2016/Presentations/Orekhov_ECLIM_poster.pdf

IOP Conf. Series: Journal of Physics: Conf. Series **907** (2017) 012019 doi:10.1088/1742-6596/907/1/012019

Comparison and analysis of the results of direct-driven targets implosion

N N Demchenko[1], G V Dolgoleva[2], S Yu Gus'kov[1,3], P A Kuchugov[1,2,4], V B Rozanov[1], R V Stepanov[1], N V Zmitrenko[2] and R A Yakhin[1]

[1] P N Lebedev Physical Institute of RAS, 53 Leninsky Av., Moscow, 119991, Russia
[2] Keldysh Institute of Applied Mathematics of RAS, 4 Miusskaya Sq., Moscow, 125047, Russia
[3] National Research Nuclear University "MEPhI", 31 Kashirskoe Av., Moscow, 115409, Russia

[4] E-mail: pkuchugov@gmail.com

Abstract. The article presents calculation results, which were received for the implosion of the typical cryogenic thermonuclear direct-drive targets that are intended for use at the OMEGA facility, NIF and Russian laser facility. The compression and burning characteristics, which were obtained using various numerical codes of different scientific groups, are compared. The data indicate good agreement between the numerical results.
Various sources of target irradiation inhomogeneity and their influence on the implosion parameters are considered. The nominal scales of these disturbances for various facilities are close to each other. The main negative effect on the efficiency of compression and burning is due to the accidental offset of the target from the center of the chamber.

1. Introduction

In recent years, considerable efforts have been made to resolve the issue that causes a discrepancy between numerical and experimental results in solving problems of laser thermonuclear fusion. This is primarily due to the failure of the National Ignition Campaign (NIC) and subsequent campaigns to achieve ignition at the National Ignition Facility (NIF). Numerous attempts by various scientific groups to explain the results have not yet been successful.

One area of research along this path is to compare the results of numerical calculations obtained using various numerical codes, analyze the differences that arise and determine possible causes for the discrepancies. This allows us to evaluate the influence of various physical processes and mathematical models, which are used for their description, on the implosion of thermonuclear targets in general. The existence and development of this trend is evidenced by the publications in recent years, which are devoted to the study of various aspects of laser thermonuclear fusion in the context of a particular laser facility.

In this paper, the authors provide and compare the modeling results of implosion of the direct-drive targets in one- and multi-dimensional formulations [intended for the OMEGA facility, NIF for the polar direct-drive (PDD) and Russian laser facility], which were obtained by the corresponding scientific groups using various numerical codes. The aim of this comparison is to determine the degree of influence of various approaches on the physical and mathematical description of the implosion process on the dynamics of compression and burning of thermonuclear fuel. In addition, possible sources of irradiation of target surface inhomogeneities and their effect on the ignition of thermonuclear fuel are analyzed. It is known that even under indirect action of the laser radiation, low-mode perturbations appear on the target, which substantially reduce the total neutron yield [1].

The given data allow one to speak about good agreement between the values of compression and burning characteristics of the capsules in various numerical calculations, as well as the similarity of the approaches used for the investigation of laser fusion problems.

2. Modeling results. Comparison and analysis

In this section, we consider typical (cryogenic) direct-drive targets and the laser pulses that are used in some operating and planned laser facilities. The results of numerical calculations that are obtained using various programs are presented.

2.1. OMEGA facility

We should start with the OMEGA facility, which implements the direct-drive approach and operates on a regular basis at the Laboratory for Laser Energetics (LLE) at the University of Rochester (USA). The total laser energy is ~26 kJ, the number of beams is 60, the operating wavelength is $\lambda_L = 0.351$ µm (Nd:Glass, 3rd harmonic of neodymium laser 3ω).To compare the numerical results, we use the target and laser pulse from [2] for which there are experimental data. Table 1 lists the values of some parameters that are available for comparison from [2].

Table 1. Results of numerical calculations of the target implosion that are intended for the OMEGA facility in a spherically symmetric formulation, which were obtained using various codes: LILAC [3], 1 – full calculation that takes into account the main processes of laser fusion problems; 2 – in contrast with the previous calculation, the Spitzer model is used to describe the heat transfer; 3 – the radiation transfer is not included in the calculation, the heat transfer is described in the diffusion approximation according to the Spitzer model; ASTER [2] – 3D calculation in a spherically symmetric formulation that takes into account the thermal conductivity according to the Spitzer model, without taking radiation transfer into account; DIANA [4] – full one-dimensional calculation; SND [5] – full one-dimensional calculation; t_c – the time of maximum compression determined from the density of DT-gas, f_{abs} – the integral fraction of the absorbed laser energy, N_n – the neutron yield.

	LILAC			ASTER	DIANA	SND	Exp.
	1	2	3				
t_c, ns	2.72	2.70	2.67	2.57	2.68	2.72	–
f_{abs}	0.542	0.536	0.540	0.480	0.542	0.542	–
N_n, 10^{14}	0.662	0.735	1.10	2.27	2.88	1.98	0.454

These data indicate good agreement between the results of various calculations, which is also confirmed by comparing the density, pressure, and temperature profiles at the time of maximum compression (see figure 1). A noticeable discrepancy in the neutron yield is due to the exponential (explosive) nature of the development of thermonuclear reactions in targets and the high sensitivity of the result to the parameters of the formed plasma.

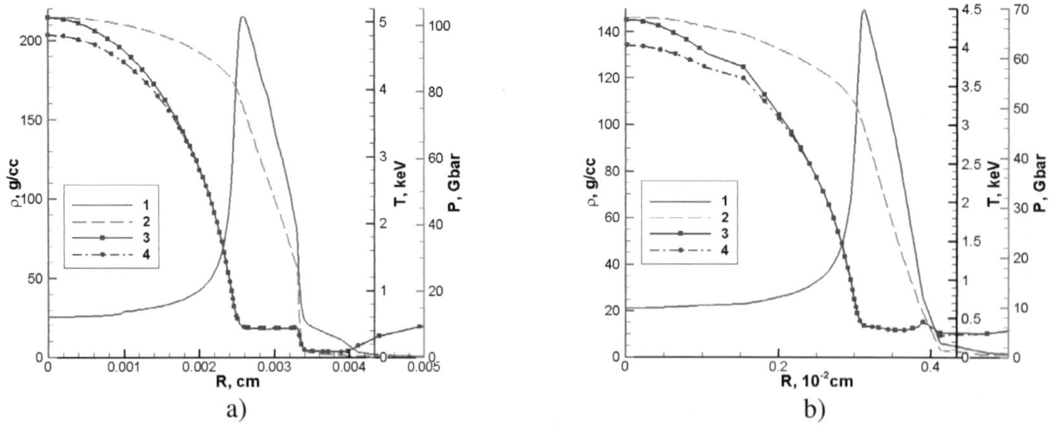

Figure 1. Profiles of the density (1), pressure (2), ion (3) and electron (4) temperatures at the moment of maximum compression of the DT-gas: DIANA code – (a) and SND – (b). See Fig. 3 in [2] to compare with the results obtained using LILAC and ASTER.

The calculation using the SND code gives the target collapse time that is closer to LILAC and demonstrates a less stressful state of the matter compared with the DIANA calculation (see table 2 and figure 1), and, as a consequence, a smaller neutron yield.

Table 2. Values of the additional characteristics of the thermonuclear target implosion, which are intended for the OMEGA facility: η_{max}, hydrodynamic efficiency; ρR, DT areal density; ρ_{max}, maximum DT density; E_r, self-radiation losses of the plasma; E_{TN}, total thermonuclear energy; $\alpha = p/p_F$, Fermi adiabat. The values η_{max}, E_r and E_{TN} are given for the moment of calculation completion, i.e., 5 ns, ρ_{max}, ρR – at the moment of maximum compression t_c.

	η_{max}, %	ρR, g/cm^2	ρ_{max}, g/cc	E_r, kJ	E_{TN}, kJ	$\alpha = p/p_F$
DIANA	10.4	0.24	212.0	0.3	0.81	3.7
SND	9.34	0.17	151.0	~0	0.56	3.3

2.2. NIF. Direct-drive targets

The failure of achieving ignition using indirect irradiation schemes led to the development of an approach for the organization of direct target irradiation using the current configuration of NIF (Nd: Glass, 3ω, 192 laser beams) - PDD [6]. This idea has significant difficulties that are associated with the peculiarities of the location of laser ports, which were originally designed to work within the framework of indirect irradiation schemes. To achieve an acceptable homogeneity of irradiation and laser energy absorption, scientific groups resort to beam repointing, which varies the total beam energy and the shape of the focal spot depending on the proximity to the equatorial region of the target. In addition, for the redistribution of energy between beams, it is proposed to use different wavelengths for the beams that have different polar angles of the optical axes that are slightly different from the nominal one (3ω) [7].

To compare the results of 1D calculations, we will use a relatively standard value of the integral fraction of laser absorption at 60%, not taking into account the specificity of PDD. Consider the characteristic cryogenic target used, for example, in [8], with the CH ablator, which can be obtained by scaling the targets that are used at the OMEGA facility to a larger amount of absorbed energy, and also resembles the baseline target for the Russian laser facility, which will be discussed in the next section. The beam parameters and dependence of the laser pulse power are given in [8]. Table 3 shows the calculations results, which are obtained using various one-dimensional numerical codes.

Table 3. Characteristics of target compression and burning, which are considered in section 2.2; G is the thermonuclear gain, V_i is the implosion velocity, and Φ is the burnout.

	G	V_i, km/s	$\alpha = p/p_F$	Φ, %
LILAC	74.6	361	2.2	15.9
DIANA	99.2	390	1.8	20.9

It can be seen that the values of various compression and burning characteristics are similar, considering the initial formulation uncertainties. The variation of the absorption fraction without changing the pulse duration leads to the fact that the value $f_{abs} = 0.85$ (0.89 without CBET, see [7]) provides the value of the gain factor of ~76 in the calculation of the DIANA program.

2.3. Russian laser facility
The total number of beams is 192, which are grouped into clusters of 4. The significant difference between the planned Russian laser facility and those considered in the previous sections is the operating wavelength of the laser radiation $\lambda_L = 0.527$ μm (Nd:Glass, 2nd harmonic of the neodymium laser 2ω). This circumstance leads to a decrease in the radiation conversion loss and, as a consequence, to a greater value of available energy that is incident on the target. In this case, the regime of laser interaction with the target material changes, and as a result, the radiation absorption decreases. Previous studies [9, 10] have shown that the transition from the 3rd to the 2nd harmonic, for the same target, leads to a noticeable decrease in the gain. Target optimization for the 2nd harmonic makes the values of the coefficient comparable to those obtained using the 3rd harmonic radiation, namely, ~20. Thus, the baseline target has been chosen, the parameters of which are given in [9]. Another important peculiarity is that the absorption occurs at a density that is less than critical, which reduces the thermonuclear yield by approximately 2 times.

3. Inhomogeneities of target irradiation at various facilities
It is known that direct-drive targets are more susceptible to the influence of hydrodynamic instabilities. Shortwave perturbations are stabilized at the ablation stage [11], and of interest are the perturbations with moderate and low modes that can be induced by the irradiation inhomogeneity or inaccuracy of positioning of a target in a chamber. The scale of various sources of inhomogeneities and the degree of their influence on the target compression at various facilities will be discussed in this section.

3.1. Scales of irradiation inhomogeneities
Let us analyze the possible sources of target irradiation inhomogeneities at various facilities. Using the data on the configuration of a laser system and the target irradiation geometry (OMEGA [2], PDD on NIF [12, 13], the Russian laser facility [10, 14]), the surface radiance of the capsule was calculated. The corresponding normalized values of intensity distributions are given in figure 2.

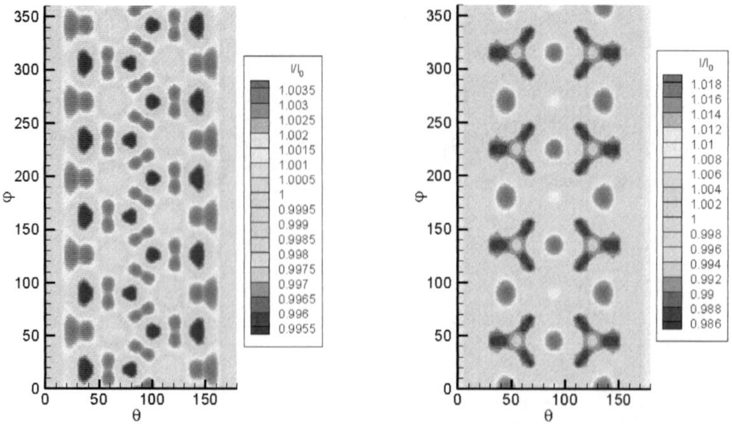

a) b)

Figure 2. Hard-sphere intensity distributions at the initial radius of targets discussed in section 2 – (a) OMEGA and (b) Russian project. For PDD at NIF see Fig. 4c in [7].

Their harmonic analysis leads to the following values of the dominant modes l_{d}: (a) $l_{\mathrm{d}} = 10$ and (b) $l_{\mathrm{d}} = 6, 8$. The value of the root-mean-square deviation of the intensity σ_0 from the mean value over the sphere, which is calculated from the distributions in figure 2, is given in table 4. It can be seen that irradiation geometry induces perturbations with moderate wavelengths and small amplitudes, which means that their influence on the compression symmetry will not be significant. Table 4 also consists of nominal values of the energy imbalance in the beams σ_{E}, beam mispointing σ_{m}, accident offset of the target from the target center σ_{s}, and beam mistiming, σ_{τ}. To take into account the possible inhomogeneities in 2D and 3D formulations, the approach that is proposed in [15] is used in this paper, except that instead of the radiance distributions, the absorbed energy in a spherically symmetric corona is used [9].

Table 4. Comparison of the scales of inhomogeneities at various laser facilities. The values for NIF are obtained by scaling the values for OMEGA ($\sim E_{\mathrm{L}}^{1/3}$). For some of them, the values from [16] are given in parentheses.

Facility	σ_0, %	σ_{E}, %	σ_{m}, μm	σ_{τ}, ps	σ_{s}, μm
OMEGA [2]	0.22	10	10	5	10
NIF	–	10(5)	40	20	40(50)
Russian laser facility	0.72	8	160	≤ 100	≤ 80

By comparing the relative values of inhomogeneities, it can be concluded that the nominal values given here are similar to each other, which ultimately should lead to similar characteristics of implosion when investigating the influence of various factors on the efficiency of the process.

3.2. Influence of irradiation inhomogeneity on the compression and burning characteristics

As was demonstrated in [2], based on the three-dimensional calculations using the ASTER code, the complex accounting of possible irradiation inhomogeneities within the nominal expected values does not make it possible to achieve a convergence of the experimental and numerical results. A consecutive individual increase in the scale of a particular source of disturbances leads to the fact that a target offset has a defining negative influence on the neutron yield, namely, a twofold increase in the displacement compared with the nominal one leads to the experimental values.

Earlier, the same conclusion was made in [17] based on a series of one- and two-dimensional calculations. For more details, see [18]. The accidental beam mispointing, as shown by numerical calculations, has the second largest effect on the efficiency of compression and burning, which is expected, because the agreed mispointing of the beams is, in fact, the target offset.

Table 5. Values of the gain (min/max) in the series of one- and two-dimensional calculations, which take into account various sources of irradiation inhomogeneities of the baseline target surface, for the Russian laser facility.

Nos.	σ_E, %	σ_m, μm	σ_s, μm	G_{1D}	G_{2D}
1	–	–	–	12.54	–
2	–	–	–	8.6/13.5	2.82/5.85
3	8	–	–	8.1/13.0	4.98/5.74
4	–	160	–	6.0/13.8	0.1/6.5
5	–	–	80	3.7/16.5	0.07/5.4

The thermonuclear gain, which is shown in table 5, illustrates the influence of certain types of irradiation inhomogeneities of a baseline target at the Russian laser facility. The first line corresponds to a spherically symmetric formulation of the problem. The second one takes into account the geometry of irradiation. The other lines take into account the geometry of irradiation and some individual sources of inhomogeneity.

4. Conclusions

The paper compares the results of numerical calculations of implosion of various thermonuclear targets in a spherically symmetric formulation. The direct impact of laser radiation on the target is a common consideration for all simulations. The comparison of corresponding numerical results for these problems obtained by different scientific groups using various numerical codes demonstrates their good agreement between themselves.

The scales of various sources of irradiation inhomogeneity, namely, the imbalance of energy in the beams, the mispointing of laser beams and errors in the positioning of the capsule inside the chamber, as well as their effect on the compression efficiency and the possibility of fusion of thermonuclear targets are analyzed. Regarding the baseline target at the Russian laser facility, it was shown that the accidental offset of the target from the center of the chamber has the most negative effect. This conclusion is consistent with the results of the analysis of the influence of inhomogeneities on the OMEGA targets, which is based on the ASTER three-dimensional code.

Acknowledgements

The research has been performed under the financial support of the Russian Foundation for Basic Research, grant No. 16-31-60101-mol_a_dk and the Program for Increasing the Competitiveness of the National Research Nuclear University "MEPhI".

The authors thank I.V. Igumenshchev for providing additional information on the calculations that are presented in the work [2].

References

[1] Clark D, Weber C, Milovich J, Salmonson J, Kritcher A, Haan S, Hammel B, Hinkel D, Hurricane O, Jones O, Marinak M, Robey P P H, Sepke S and Edwards M 2016 *Physics of Plasmas* **23** 0563022

[2] Igumenshchev I V, Goncharov V N, Marshall F J, Knauer J P, Campbell E M, Forrest C J, Froula D H, Glebov V Y, McCrory R L, Regan S P, Sangster T C, Skupsky S and Stoeckl C 2016 *Physics of Plasmas* **23** 052702

[3] Delettrez J, Epstein R, Richardson M C, Jaanimagi P A and Henke B L 1987 *Physical Review A*

36 3926-34

[4] Zmitrenko N, Karpov V, Fadeev A, Shelaputin I and Shpatakovskaya G 1983 *Voprosy Atomnoy Nauki i Tekhniki (VANT) Series Methods and Software for Numerical Solution of Problems of MathematicalPhysics* 34-7

[5] Dolgoleva G 1983 *Voprosy Atomnoy Nauki i Tekhniki (VANT) Series Methods and Software for NumericalSolution of Problems of Mathematical Physics* 29-33

[6] Skupsky S, Marozas J A, Craxton R S, Betti R, Collins T J B, Delettrez J A, Goncharov V N, McKentyP W, Radha P B, Boehly T R, Knauer J P, Marshall F J, Harding D R, Kilkenny J D, Meyerhofer D D,Sangster T C and McCrory R L 2004 *Physics of Plasmas* **11** 2763-70

[7] Hohenberger M, Radha P B, Myatt J F, LePape S, Marozas J A, Marshall F J, Michel D T, Regan S P, SekaW, Shvydky A, Sangster T C, Bates J W, Betti R, Boehly T R, Bonino M J, Casey D T, Collins T J B,Craxton R S, Delettrez J A, Edgell D H, Epstein R, Fiksel G, Fitzsimmons P, Frenje J A, Froula D H,Goncharov V N, Harding D R, Kalantar D H, Karasik M, Kessler T J, Kilkenny J D, Knauer J P, Kurz C,Lafon M, LaFortune K N, MacGowan B J, Mackinnon A J, MacPhee A G, McCrory R L, McKenty P W,Meeker J F, Meyerhofer D D, Nagel S R, Nikroo A, Obenschain S, Petrasso R D, Ralph J E, RinderknechtH G, Rosenberg M J, Schmitt A J, Wallace R J, Weaver J, Widmayer C, Skupsky S, Solodov A A, Stoeckl C, Yaakobi B and Zuegel J D 2015 *Physics of Plasmas* **22** 056308

[8] Lafon M, Betti R, Anderson K S, Collins T J B, Epstein R, McKenty P W, Myatt J F, Shvydky A and Skupsky S 2015 *Physics of Plasmas* **22** 032703

[9] Bel'kov S A, Bondarenko S V, Vergunova G A, Garanin S G, Gus'kov S Y, Demchenko N N, Doskoch I Y,Kuchugov P A, Zmitrenko N V, Rozanov V B, Stepanov R V and Yakhin R A 2015 *Journal of Experimentaland Theoretical Physics* **121** 686-98

[10] Rozanov V B, Gus'kov S Y, Vergunova G A, Demchenko N N, Stepanov R V, Doskoch I Y, Yakhin R A andZmitrenko N V 2016 *Journal of Physics: Conference Series* **688** 012095

[11] Takabe H, Mima K, Montierth L and Morse R 1985 *Physics of Fluids* **28** 3676

[12] Collins T J B, Marozas J A, Anderson K S, Betti R, Craxton R S, Delettrez J A, Goncharov V N, HardingD R, Marshall F J, McCrory R L, Meyerhofer D D, McKenty P W, Radha P B, Shvydky A, Skupsky Sand Zuegel J D 2012 *Physics of Plasmas* **19** 056308

[13] Weilacher F, Radha P B, Collins T J B and Marozas J A 2015 *Physics of Plasmas* **22** 032701

[14] Garanin S, Bel'kov S and Bondarenko S 2012 Conception of UFL-2M laser facilitiy Book of Abstrats of the39th Zvenigorod Int. Conf. on Plasma Physics and Controlled Fusion, Zvenigorod, Russia, 6-10 February2012

[15] Gus'kov S, Demchenko N, Zhidkov N, Zmitrenko N, Litvin D, Rozanov V, Stepanov R, Suslov N and YakhinR 2010 *Journal of Experimental and Theoretical Physics* **111** 466-83

[16] Eimerl D, Rothenberg J, Key M, Weber S, Verdon C, Skupsky S, Sources J and Craxton S 1995 ConfiguringNIF for Direct Drive Experiments Preprint UCRL-JC-121271 Lawrence Livermore National Laboratory

[17] Demchenko N N, Gus'kov S Y, Kuchugov P A, Rozanov V B, Stepanov R V, Vergunova G A, Yakhin R Aand Zmitrenko N V 2016 *Journal of Physics: Conference Series* **717** 012032

[18] Bel'kov S A, Bondarenko S V, Vergunova G A, Garanin S G, Gus'kov S Y, Demchenko N N, Doskoch I Y,Kuchugov P A, Zmitrenko N V, Rozanov V B, Stepanov R V and Yakhin R A 2017 *Journal of Experimentaland Theoretical Physics* **124** 341-51

IOP Conf. Series: Journal of Physics: Conf. Series **907** (2017) 012020 doi:10.1088/1742-6596/907/1/012020

Hollow poly(alpha-methylstyrene) shells for inertial confinement fusion targets

A V Pastukhov[1,2,3], V A Davankov[2], A A Akunets[1], N G Borisenko[1], A S Orekhov[1], S M Tolokonnikov[2] and K S Pervakov[1]

[1] P N Lebedev Physical Institute of the Russian Academy of Sciences, 53 Leninskiy Prospekt, Moscow 119991, Russia
[2] A N Nesmeyanov Institute of Organoelement Compounds of the Russian Academy of Sciences, 28 Vavilova str., Moscow 119991, Russia

[3] E-mail: avpast@gmail.com

Abstract. Using a low-temperature cationic polymerization technique, a number of poly(alpha-methylstyrene) samples with molecular masses of 300000–400000 were prepared, and their structural parameters were examined via nuclear magnetic resonance and X-ray diffraction techniques. It was found that all synthesized polymers display a stereoregular syndiotactic structure of macro chains and are amorphous. Using these polymers and the microencapsulation technique, hollow shells were obtained that were 2.4 mm in diameter and had a wall thickness of 10 to 60 μm. The quality of walls and surfaces of the shells was analyzed using scanning electron microscopy and optical profilometry. The hollow shells are intended for the preparation of inertial confinement fusion targets and for the laser plasma research.

1. Introduction

Preparation of high-quality spherical hollow shells is one of the most important technical problems in the development of targets for the Inertial Confinement Fusion (ICF) systems. Currently, spherical beryllium hollow shells that meet very strict requirements with respect to their geometry and surface quality are considered as the most promising approach for preparing useful internal confinement targets. The preparation of these metal shells involves two major steps: deposition of metal vapors on the surface of polymeric shells and thermal elimination of the initial polymeric material. Accordingly, one must first develop a technique for preparing polymeric shells with a perfect spherical shape and a defect-free surface. Poly(alpha-methylstyrene) (PAMS) has been found [1–3] to be one of the most useful polymers. Still, it is characteristic of all polymeric materials that every batch of PAMS will exhibit its individual physical properties depending on the synthesis conditions (nature of the catalyst, temperature and other process parameters). Nevertheless, the specificity of the final targets is determined by the polymer precursor used for the preparation of hollow shells to fit into a well-defined range of thermal and mechanical properties. Thus, the polymer must degrade completely with the evolution of only volatile products during the thermal treatment at 250–300 °C, while the thin walls of the polymer shell (10–50 μm) must exhibit sufficient rigidity and withstand the process of elimination of the "internal" water that initially fills the shell.

Accordingly, the present work is aimed at the synthesis of several batches of PAMS by varying the synthesis conditions, optimization of parameters of the droplet formation stage of the fabrication process of thin-wall hollow shells composed of the obtained polymers and, finally, examination of the main quality characteristics of the shells.

2. Experimental setup

High-molecular-weight PAMS is known as a useful starting material for the preparation of Inertial Confinement Fusion Targets [4–6]. To produce the material of required quality, the low-temperature cationic polymerization approach was chosen [**7**]. Examination of the obtained polymer structure was performed using proton nuclear magnetic resonance (NMR) and X-ray diffraction techniques. Proton magnetic resonance spectra were recorded in a deuterated chloroform solution of the PAMS samples (at a concentration of 1%) using a Bruker AV 400 instrument (400.13 MHz for ^{1}H). Diffraction curves of the pressed at 205 °C and 1000 atm PAMS samples were recorded using a Rigaku MiniFlex 600 diffractometer with a copper cathode in the range of 2θ angles from 3 to 150 at a rate of 0.3 degree/min. The final hollow shells, which were prepared from the polymer batches, were examined using scanning electron microscopy (SEM) and optical profilometry techniques. Micrographs of the shell surface and fracture chips were obtained using an SEM instrument Jeol JSM-7001F with a Schottky cathode. Profiles of the shell surface were examined using a Zygo NewView profilometer equipped with a 10× micro interferometer.

3. Results and Discussion

The choice of cationic polymerization approach for high molecular weight PAMS is justified by a lower sensibility of this method to eventual contaminations in the starting monomer, compared with anionic polymerization. By varying the synthesis conditions, a number of PAMS batches were obtained that differed from each other in chain structure and physical properties, mainly in the molecular weight distribution characteristics (molecular weight from 300000 to 400000). In particular, the samples PAMS20, 25, 30, 32 were obtained using boron trifluoride as a catalyst (PAMS31 – with boron trichloride) at –70 to –76 °C (PAMS31 –80 °C). The PAMS35 sample was obtained by catalyzing anionic polymerization with butyl lithium at –79 to –85 °C.

The examination of the synthesized polymer structures using the X-ray structure analysis technique (figure 1, a) resulted in a conclusion that all materials are amorphous because their diffraction curves display only two reflexes, which is characteristic of amorphous polymers of this type.

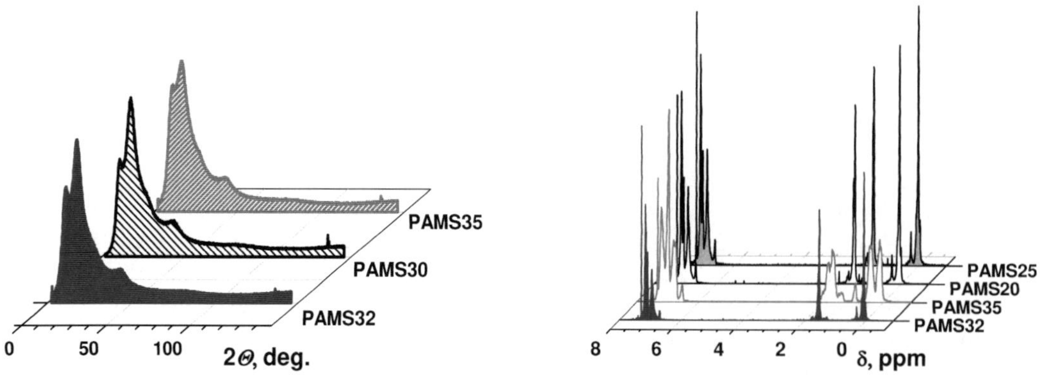

a) b)

Figure 1. Results of structural investigations of the synthesized polymers, (a) diffraction curves and (b) ^{1}H NMR spectra.

Besides, the ^{1}H NMR spectra of all samples, which were obtained using the cationic polymerization approach (figure 1b), reveal a resonance signal of methyl protons near 0.3 ppm, which points to the syndiotactic arrangement of monomer units in the PAMS chain [8]. Thus, these polymers are stereoregular. This is contrary to atactic PAMS, which results from anionic polymerization [7]. The NMR spectrum of the PAMS35 sample of that type is characteristic of atactic structure of polymer chains [9].

Hollow shells that were produced from the synthesized PAMS samples were prepared using the microencapsulation technique [10]. Solutions of the polymers, which were injected into a special system of capillaries, had 10 and 20% concentrations and were based on various solvents: benzene, toluene, ortho-xylene, ethylene dichloride, tetrachloroethane, dichloropropane, fluorobenzene, and fluorotoluene. The aqueous phase, which was used as a medium to harden polymer shells by removing the solvent during the careful heating of the mixture, contained polyvinyl alcohol at a concentration of 1.5–2.0% and various ammonium salts at a similar concentration of 1–2%. The hardening of the shell droplets was carried out under a number of combinations of temperature and time conditions. Thus, the examination of solid shells, which was obtained using a profilometer, revealed a substantial difference in the surface topology and wall-forming polymer structure, which depended on the preparation protocols (figure 2).

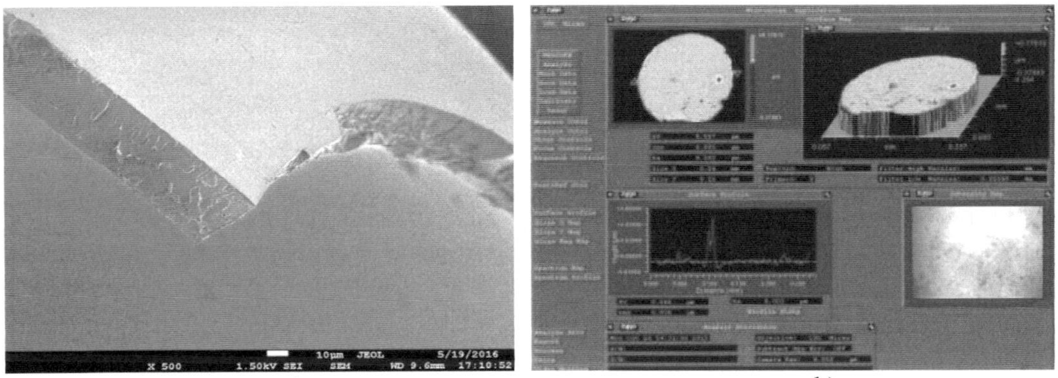

a) b)

Figure 2. SEM micrograph of the shell fracture (a) and profilometer study of the shell surface (b) of the product PAMS30.

The shell-wall cross-section was studied to obtain uniform spherical shells. The surface of the cut edge revealed a different character of the local mechanical destruction of the polymer phase. Specifically, we came across either a ductile or a brittle character for different routes. Some shell batches have small inclusions of ammonium nitrate (as measured via Energy-dispersive X-ray spectroscopy) in the polymer layer or on its surface if ammonium nitrate was present in the aqueous phase. SEM micrographs of other batches display bubbles of various sizes in the polymer shell wall. On the surface of shells, these bubbles may form craters of different depths or distort the surface in one or another way if the bubbles remain close to the surface. There are fabrication routes that reveal a higher bubble concentration and foam structure of the wall.

Shells that have thin walls, less than 10–20 µm, had the smoothest surface. However, the profilometer and micrography measurements of the entire shell demonstrated that shells with very thin walls, less than 10 µm, were mostly non-spherical, rugby ball-shaped. The smallest deviation from the regular spherical form was characteristic of the shells with thick walls, approximately 50–70 µm. Regardless of the aims of the present work, the rugby ball-shaped shells can find their use in the non-symmetric irradiation schemes and for particle source optimization.

The process of shell hardening that freezes the geometric form of the shell droplet is governed by a superposition of several factors. Specifically, these factors are the rate of diffusion of solvent molecules from the polymer solution into the aqueous environment media, rate of concomitant changes of the rheological properties of the polymer phase of the walls, variation of the surface tension of the two-phase system, deformational resistance of the hardening droplet shell and some other factors. Therefore, it was necessary to conduct many experiments, where different compositions of initial polymer solutions and aqueous phase, temperature and time allowed for the hardening of the shells were systematically examined. Finally, rigorous optimization of the preparation protocols allowed to obtain shells with a regular spherical form and high-quality surfaces (figure 3).

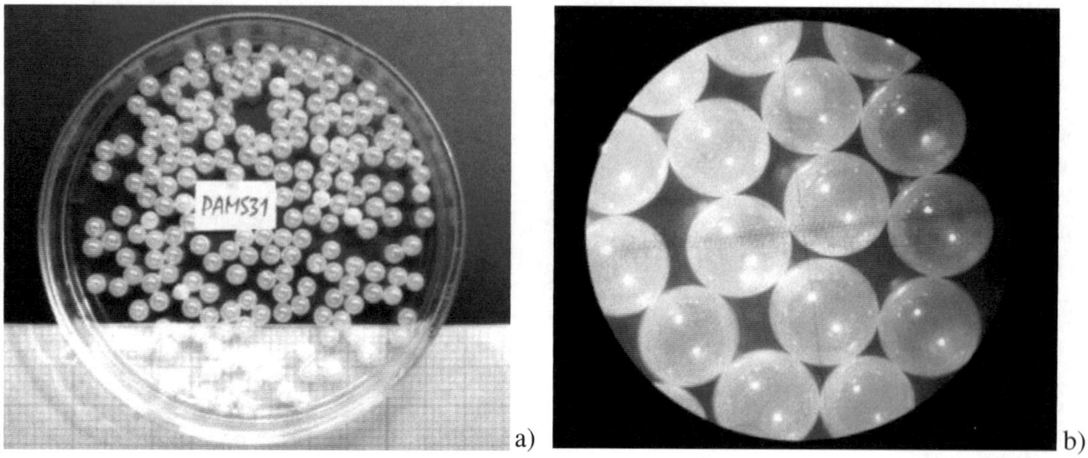

Figure 3. General view of hollow shells prepared from poly(alpha-methylstyrene) PAMS31 (a), and the microphotograph of these shells at 16× (b).

4. Conclusions

Based on poly(alpha-methylstyrene) with a molecular mass of 300000–400000 and using low temperature cationic polymerization of the monomer, hollow shells were obtained for the further use in ICF targets. By systematically varying the polymer synthesis conditions and technical protocols of the microencapsulation process, hollow spherical shells were obtained with a diameter of up to 2.4 mm and a shell thickness of 10 to 60 μm. The rugby ball-shaped shells can also find their possible applications in laser plasma research.

Acknowledgments

Physical properties of the obtained material were examined using the equipment offered by the LPI Shared Facility Centre. The work was partially supported by the RFBR grant N15-52-45116.

References

[1] Cook R, McQuillan B, Takagi M and Stephens R 2000 *Inertial Confinement Fusion Semiannual Report* **1** 1
[2] McQuillan B W, Greenwood A 1999 *Fusion Technology* **35** 194
[3] Cook R, Buckley S R, Fearon E and Letts S A 1999 *Fusion Technology* **35** 206
[4] Koroskenyi B, Wang L and Faust R 1997 *Macromolecules* **30** 7667
[5] Li D, Hadjikyriacou S and Faust R 1996 *Macromolecules* **29** 6061
[6] Okamura S, Higashimura T and Imanishi Y 1958 *J. Pol. Sci.* **126** 491
[7] Braun D, Cherdron H and Kern W 1971 *Praktikum der Makromolekularen Organischen Chemie 2nd* (Hüthig Verlag, Heidelberg)
[8] Toman L, Pokorny S and Spevacek J 1989 *J. Pol. Sci. A* **27** 2217
[9] Berger P A, Kotyk J J and Remsen E E 1992 *Macromolecules* **25** 7227
[10] Takagi M, Cook R, McQuillan B, Elsner F, Stephens R, Nikroo A, Gibson J and Paguio S 2002 Preparation of NIF scale poly(alpha-methylstyrene) mandrels *Lawrence Livermore National Laboratory Report UCRL-ID-148920*

IOP Conf. Series: Journal of Physics: Conf. Series **907** (2017) 012021 doi:10.1088/1742-6596/907/1/012021

Dynamics of impulse volume discharge formation in atmospheric pressure helium

V S Kurbanismailov[1], O A Omarov[1], G B Ragimkhanov[1,3] and D V Tershonok[2]

[1] Dagestan State University, M Gadjieva st. 43a, Makhachkala, Russia 367003
[2] Joint Institute for High Temperatures of RAS, Izhorskaya st. 13 Bd.2, Moscow, Russia 125412

[3] E-mail: gb-r@mail.ru

Abstract. In the article, we present the results of investigation of the initial stage of volume electric discharge formation in atmospheric pressure helium in inhomogeneous pre-ionized gas. The peculiarities of cathode-directed ionization wave development between two flat electrodes in atmospheric pressure helium are investigated.

1. Introduction

Nonequilibrium and unstable plasma of impulse volume electric discharge can be widely used in power gas lasers, optical sources, etc. Despite a number of articles that are devoted to impulse volume discharges, many issues that are connected with physics of impulse breakdown and mechanisms of the initial stages formation are still not fully investigated and provoke scientific discussions [1–6].

Therefore, it is undoubtedly interesting to investigate the experimental and theoretical data of consistent dynamics of the impulse volume discharge formation and development during the preparatory gas ionization in helium (He) at atmospheric pressure.

2. Methods of investigation

2.1. Experimental setup and methods of investigation

The experimental setup and methods of investigation have been presented in [7]. The voltage in the gaps changes within the limits of 3÷20 kV. The duration of electric pulse is determined by the discharge time of the charged capacitor, and the characteristic discharge time of the capacitor under the experimental conditions was $\tau \geq 10^{-5}$. The space-time discharge development was recorded using a photoelectronic register (PER2-1). Preparatory gas ionization ($n_0 \sim 10^8$ cm^{-3}) was obtained via gap radiation using a mesh anode via an exterior spark discharge UV radiation from a third-party discharge, which was located on the side at a distance of 5-7 cm from the axis of the main electrodes in the same gas [8]. The investigated gap (1-cm length) was formed via solid aluminium electrodes with a diameter of 4 cm.

2.2. Formation model of impulse volume discharge in helium at atmospheric pressure.

Currently, there are three approaches for the numerical simulation of plasma: kinetic (in which Boltzmann equations are numerically solved either directly or using the microparticle method with respect to collisions according to the Monte Carlo method [9, 10]), hydrodynamical (in which the equations are solved numerically for the first several moments of Boltzmann equation [10, 11]), and a

IOP Conf. Series: Journal of Physics: Conf. Series **907** (2017) 012021 doi:10.1088/1742-6596/907/1/012021

hybrid method [12]. The choice of the model is defined by the free path length for every type of the particles. For discharge under the atmospheric-pressure, the hydrodynamical and hybrid methods are usually used for numerical simulation.

In this article, to describe the discharge, we use the 2D axial-symmetric drift-diffusion model of electron and ion movements along with the Poisson equation [13–16]:

$$\frac{\partial n_e}{\partial t} + \nabla \cdot \vec{\Gamma}_e = \alpha_{nl}\left|\vec{\Gamma}_e\right| - \beta n_e n_i,$$

$$\frac{\partial n_i}{\partial t} + \nabla \cdot \vec{\Gamma}_i = \alpha_{nl}\left|\vec{\Gamma}_e\right| - \beta n_e n_i,$$

$$\nabla \cdot \vec{E} = \frac{e}{\varepsilon_0}(n_+ - n_e)$$

Here, the electron and ion fluxes can be expressed as $\vec{\Gamma}_e = -n_e\mu_e\vec{E} - D_e\nabla n_e$ and $\vec{\Gamma}_+ = n_+\mu_+\vec{E} - D_+\nabla n_+$, respectively, where $\vec{j}$ is the current density; n_e and n_+ are the density of electrons and ions, respectively; α_{nl} is the nonlocal first Townsend coefficient, and β is the recombination coefficient; $\mu_e(E/N)$ and $\mu_+(E/N)$ are the electron and ion mobility, respectively; $D_e = \frac{kT_e\mu_e}{e}$ and $D_+ = \frac{kT_+\mu_+}{e}$ are the electron and ion diffusion coefficients, respectively; ε_0 is the dielectric constant; $\vec{E}$ is the electric field strength; φ is the electric potential; N is the concentration of neutral particles; $T_e(E/N)$ is the electron temperature; $T_+ = T_0$ is the ion temperature; e is the elementary charge. Transport coefficients, ionization and deionization parameters are obtained from the dependences in [14, 15].

The non-local ionization coefficient was obtained from [16].

The densities of electrons and ions at the initial time inside the computational domain are: $n_e(x, y, 0) = n_+(x, y, 0) = n_0$. (x - dimension in radial direction).

The boundary conditions for the cathode are $y = 1$ cm, $\vec{\Gamma}_e = \gamma\vec{\Gamma}_+$, $\partial n_+/\partial y = 0$, $\varphi_k = 0$, where $\gamma = 0{,}1$ is the secondary emission ratio, and for the anode, $y = 0$ cm,

$$\partial n_e/\partial y = 0,\ n_+ = 0,\quad \varphi_k = U(t) - I(t)R_b,\ I(t) = 2\pi\int_0^r j_a(x,t)x\,dx,\ \partial\varphi/\partial y = 0.$$

where $I(t)$ is the total current in the electrical circuit and $j_a(x,t)$ is the distribution of current density on the anode.

The pre-ionization electron and ion concentrations in the gap are set as follows:

$$n_e(x, y) = n_{e0}\cdot\frac{\exp(-k\cdot x^2)}{r^2}\cdot\frac{\exp(-k\cdot y^2)}{d^2};\quad n_i(x, y) = n_{e0}\cdot\frac{\exp(-k\cdot x^2)}{r^2}\cdot\frac{\exp(-k\cdot y^2)}{d^2},$$

where $r = 2$ cm is the discharge gap radius, $d = 1$ cm is the interelectrode distance, and k is the coefficient that characterizes the level of inhomogeneous pre-ionization [$k = (1\div10)$].

3. Results of the research and their investigation

3.1. Results of the experimental research and their investigation.
The most interesting are the experimental results of direct observation of the discharge formation dynamics with space and time resolution in the nanosecond time range, which were obtained using PER2-1.

In particular, in fig. 1 (photos *1-4*), the photos illustrate the volume discharge formation in He at different time periods. The discharges are made via preliminary gas ionization.

As can be seen in fig. 1 (photos *1–4*), the first registered glow appeared on the anode after the outer field application, which then spreads to the cathode in the form of diffuse glow with a characteristic speed of ~10^7 cm/s. The glow front is inhomogeneous, and the intensity falls from the axis of discharge to the periphery. This shows the higher intensity of ionized processes on the discharge axis. After the glow front arrives to the cathode (photo *1–3*), the discharge proceeds to the next phase – the overall combustion phase.

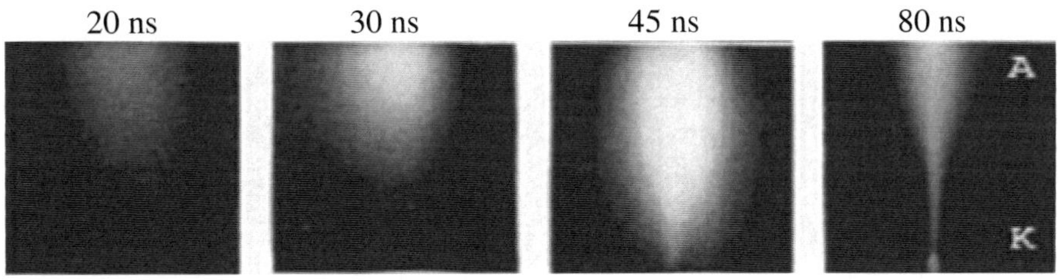

Figure 1. Dynamics of volume discharge formation in He (photos *1–4*) and the integral picture of discharge glow (photo *5*): $p = 760$ torr, $d = 1$ cm, $U_0 = 9$ kV.

For a small breakdown voltage $U_0 < 6$ kV, a volume discharge with a high glow homogeneity and combustion duration is formed. When the voltage is $U_0 > 6$ kV (look at fig. 1, photo *4*) and the current density is $j \geq 40$ A/cm^2, the plasma channels, which are connected with cathode spots, are formed in the pre-cathode area.

3.2. Results of the numerical experiment and their consideration.
Let us look at the results of numerical experiment of the discharge formation stage in inhomogeneous gas preionization. As one can see from the results, after the high-tension impulse is applied, the electrons due to drift move to anode from the cathode region. Simultaneously, the electron concentration in the cathode region decreases, and their concentration in the discharge column increases due to ionized electron multiplication. The appearance of uncompensated positive electron cloud (look at fig. 2a) makes the field in the cathode region stronger and, at the same time, weaker in the positive column region (look at fig. 2b).

Gas preionization inhomogeneity due to both far and wide discharge gap leads to the inhomogeneity of electron concentration distribution when the discharge gap is wide. Thus, the ionization wave front is not flat, it has an oblong shape. The ionization wave formation mostly occurs in the discharge gap axis, where the field has the maximum value.

In the ionization wave development process, the uncompensated positive discharge on the front creates a high transverse electric field E_x. When it approaches the cathode, the field tension on the ionization front also increases, similar to the intensity of ionization processes. When there is a decrease in the preionization electron concentration, the time of the ionization wave formation increases as well as the discharge formation time.

In the formed plasma column, the concentration of charged particles on the discharge axis is higher, and it becomes smaller at the discharge periphery.

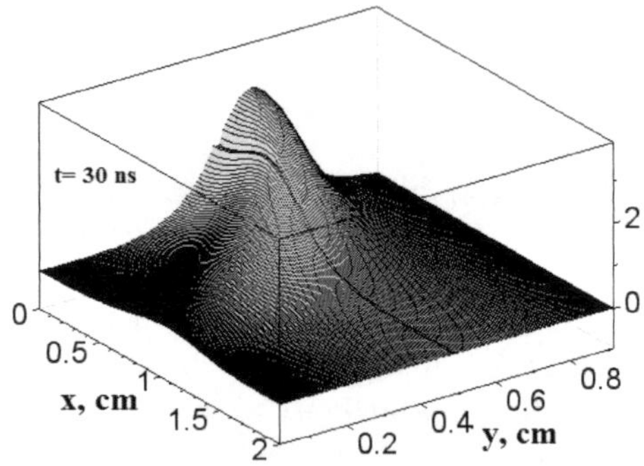

Figure 2 a. Concentration distribution n_i- n_e, 10^{10} [cm^{-3}] in the interelectrode gap: U_0 = 9 kV, p = 760 torr, d = 1 cm, t = 30 ns.

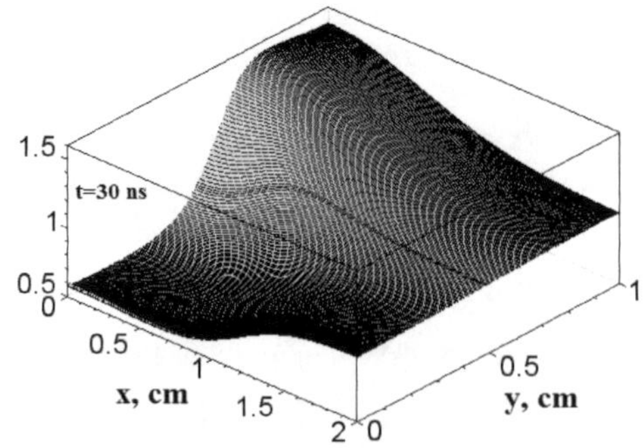

Figure 2 b. Direct-axis component of the field tension Ey, 10^4 [V/cm] in the interelectrode gap: U_0 = 9 kV, p = 760 torr, d = 1 cm, t = 30 ns.

4. Conclusions

Discharge formation occurs during the propagation of cathode ionization wave. This is proven by the results of numerical experiments and by the results of experimental research of the space-time dynamics of the discharge initial stages development. Wide gas preionization homogeneity forms inhomogeneous ionization wave front, which is developed from the central zone of the gap.

The studies that are performed in this work show the complexity of the processes in the discharge formation stages. This generates interest for the further study of the problem.

Acknowledgments

The theoretical part was supported by the President of Russia Scholarship for young scientists No. 3812.2016.1.

References

[1] Tarasenko V, Baksht E and Burachenko A 2010 Technical Physics Letters. **36**(4) 375-78.
[2] Naidis G and Walsh J 2013 *J. Phys. D: Appl. Phys.* **46** (095203) 13
[3] Tarasenko V and Yakovlenko S 2004. *Phys. Usp.* **47**(887)
[4] Babich L, Bochkov E and Kutsyk I 2014 *JETP Letters* **99**(7) 386–90
[5] Osipov V V 2000 *Physics-Uspekhi* **170**(3) 225-245
[6] Kurbanismailov V, Omarov O, Ragimkhanov G et al 2016 *Plasma Physics Reports* **42**(7) 687-698
[7] Kurbanismailov V and Omarov O 1995 *High Temperature* **3**(3) 365-369
[8] Kurbanismailov V, Omarov O and Khachalov M 1989 *Measurement technology* **3** 30-37
[9] Birdsall C K 1991 IEEE Trans. Plasma Sci..**19**(2) 65-85
[10] Kim H C, Yang S S, etal. 2005 *J. Phys.* D: *Appl. Phys.* **38** 283-301
[11] Georghiou G E, Papadakis A P, Morrow R A, and Metaxas A C 2005 *J. Phys.* D: *Appl. Phys.* **38** 303–328
[12] Kushner M J 2005 *J. Phys.* D: *Appl. Phys.* 38 1633-1643
[13] Tereshonok D 2014 *Technical Physics Letters* **40**(3) 83-89
[14] Raiser U 2009 *Physics of gas discarge* MFTI. Dolgoprudny: Intellect 736
[15] Surzhikov S 2006 *Physical mechanics of the gas discharges* M.: Publishing house MSTU under the name of N.E. Bauman 640
[16] Soloviev V and Krivtsov V 2009 *J. Phys.* D: *Appl. Phys.* **42** 13

IOP Conf. Series: Journal of Physics: Conf. Series **907** (2017) 012022 doi:10.1088/1742-6596/907/1/012022

Subthreshold self-sustained discharge initiated by a microwave beam in a large volume of high-pressure gas

K V Artem'ev[1], G M Batanov[1], N K Berezhetskaya[1], A M Davydov[1], I A Kossyi[1,3], V I Nefedov[2], K A Sarksyan[1] and N K Kharchev[1]

[1] Prokhorov General Physics Institute, Russian Academy of Sciences, Moscow, 119991 Russian Federation
[2] Moscow State University of Information Technologies, Radio Engineering, and Electronics, Moscow, 119454 Russian Federation

[3] E-mail: kossyi@fpl.gpi.ru

Abstract. The presented results are based on the experimental studies of generating a long plasma column in gas (or gas mixtures) at atmospheric pressure both in free space and in a closed chamber. The microwave generator GYROTRON was used as an energy source. Its parameters were as follows: the microwave pulse power was in the range of $200 \leq P \leq 600$ kW, the wavelength was $\lambda \approx 0.4$ cm, and the pulse duration was in the range of $0.5 \leq \tau i \leq 20$ ms. Under strong subthreshold conditions, a plasma column with a length of up to $L \leq 50$ cm was created using a microwave beam that was formed using a quasi-optical mirror system. The discharge initiation system had an original design. Based on the discharge structure, expansion dynamics and typical discharge plasma parameters, this discharge can be attributed to the type of microwave discharges that are known as self/non-self-sustained discharges. The discharge properties and advantages of using this discharge as a basis of a plasma-chemical reactor are discussed.

1. Introduction

Studies of the fundamental physical problems of microwave-driven discharges in gases, which are carried out at the Prokhorov General Physics Institute, have provided the basis for the development of a series of original plasma sources that are intended for the use both in laboratory experiments and for solving modern technological problems. The most promising plasma source from the technological point of view is the plasmatron, which is based on the subthreshold discharge that is initiated by a pulsed microwave beam [1, 2]. This new type of microwave discharge, which was named as a self/non-self-sustained (SNSS) discharge [3–6], occurs when the following initial conditions are fulfilled:

(i) Pressure of the gas in which the microwave beam propagates should be high so that the strong inequality is true:

$$\nu_{eff} \gg \omega \qquad (1)$$

(here, ν_{eff} is the effective frequency of electron-neutral collisions, and ω is the microwave frequency);

(ii) Initially, the reduced microwave electric field should be low so that a self-sustained discharge could not be excited anywhere over the beam volume:

$$\gamma \equiv E_0/n_m << [E_0/n_m]_{thr} \qquad (2)$$

(here, E_0 is the amplitude of the microwave electric field, n_m is the density of neutrals, and $[E_0/n_m]_{thr}$ is the threshold reduced electric field that is needed for the initiation of a self-sustained discharge);

(iii) There should be a discharge initiator installed in a volume V_d, which is much smaller than the microwave beam volume V_{MW}, so that

$$V_d << V_{MW}, \qquad (3)$$

and the initiation time τ_i should be shorter (or much shorter) than the duration of the microwave pulse τ_{MW}:

$$\tau_i << \tau_{MW} \qquad (4)$$

Under the conditions when the microwave electric field is smaller (or much smaller) than the breakdown field, the discharge may be excited locally inside the region occupied by the microwave beam where either the electric field of the electromagnetic wave increases locally or the density n_m decreases so that relation (2) becomes reversed. The schematic sketch for the three versions of initiation of an SNSS-discharge, which is used at the (GPI General Physics Institute), is presented in Figure 1.

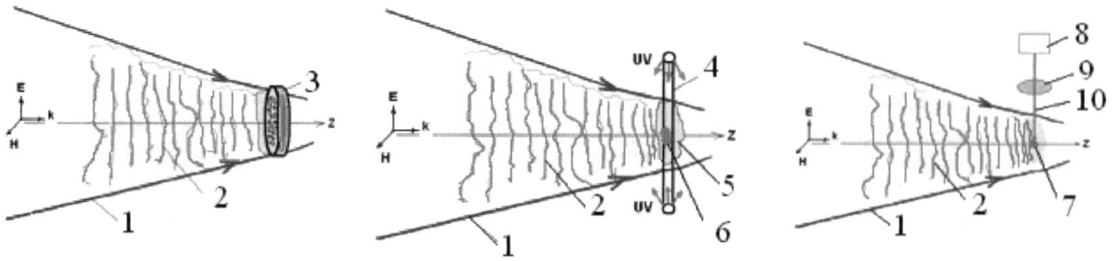

Figure 1. Schemes of the SNSS-discharge initiation. a) Metal-dielectric target; b) Ring-shaped source of UV-radiation; c) CO_2 laser spark. 1-microwave beam; 2-SNSS-discharge; 3-metal-dielectric initiator; 4-annular multispark initiator; 5-non-self-sustained discharge that initiates the SNSS-discharge; 6-first local microplasma formation; 7-CO_2 laser spark; 8-CO_2 laser; 9-lense; 10-CO_2 laser beam.

Thus, metal-dielectric targets were installed at some cross section for the initiation of an extremely "subthreshold" self-sustained discharge in [6] (scheme figure 1a). In [7], to initiate the excitation of a localized discharge in the microwave beam, we used a CO_2 laser spark (scheme in figure 1b). Figure 1c shows a schematic of the experiment, in which the SNSS-discharge was initiated using a ring multi-spark source of UV radiation, which produces photo-ionization plasma, namely the non-self-sustained microwave discharge. Eventually, its instability may result in the self-sustained plasma stage near the axis because bright filaments appear along the microwave beam electric field [8, 9].

All of the initiation methods in our experiments, which were conducted at the GPI, demonstrated that the local excitation of a self-sustained discharge in a volume that is much smaller than the volume occupied by the microwave beam is accompanied by an ionization wave. This wave propagates toward the microwave generator and builds an axial plasma column. Therefore, when the beam pulse duration is larger, the plasma column length is larger. Thus, we unexpectedly observed a new phenomenon, namely, the formation of an extended gas-discharge plasma column in microwave fields that are smaller (and even substantially smaller) than the threshold fields for exciting a self-sustained discharge in gas in free space. The characteristic photographs of SNSS-discharges, which are initiated using metal-dielectric targets and a CO_2 laser spark, are presented in figures 2a and 2b.

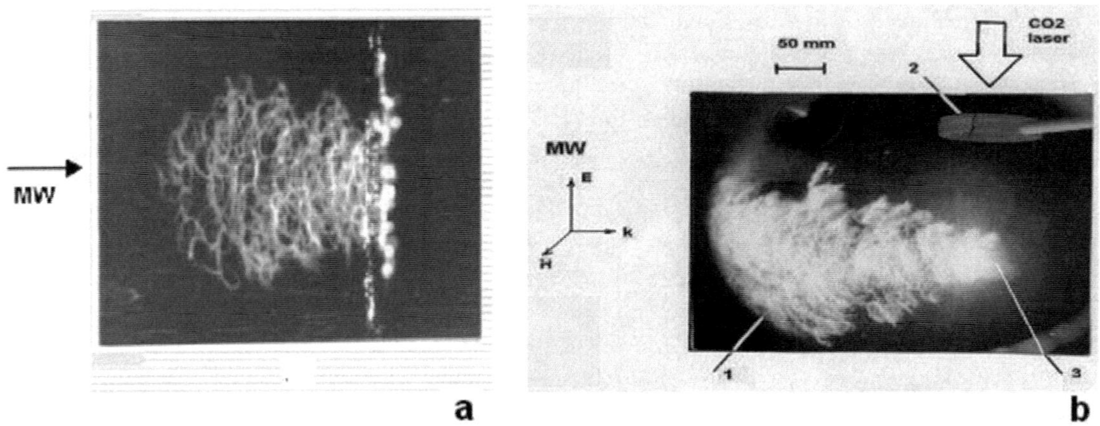

Figure 2. a) Photograph of the SNSS-discharge excited in air using a subthreshold microwave beam via the metal-dielectric target (p = 200 Tor; λ = 2.5 cm); b) Photograph of the SNSS-discharge excited in air using a subthreshold microwave beam through the initiation via the CO_2 laser spark (p = 200 Tor; λ = 2.5 cm).

Note that all known experimental data on SNSS-discharges were obtained under the following conditions:

- Gas pressure below the atmospheric pressure ($p \leq 200$ Torr);
- Centimeter wavelengths of microwave radiation ($\lambda \approx 2$–5 cm);
- Moderate microwave pulse power ($P \leq 100$ kW).

The purpose of the present work was to extend the area of fundamental studies of the discharge: to shift the pressure range toward values that are close to (and even somewhat larger than) the atmospheric pressure, to extend the range of pulsed microwave beam powers up to $P \approx 1$ MW, and to switch to the millimeter range of the microwaves ($\lambda \approx 0.2$–0.8 cm). The recent interest in the realization of these ranges of parameters – gas and microwave beam – is motivated by our firm intention to advance toward the construction of an adequate physical model of the gas-discharge phenomenon, as well as, by our interest in promising plasmachemical applications.

The present paper is devoted to a brief analysis of the physical processes underlying this unusual microwave gas-discharge formation. The pioneering experiments and research on SNSS-discharges in gases at atmospheric pressure, which have been conducted at the GPI using a high-power pulsed microwave beam in the millimeter wavelength range, are described below. The properties and advantages of the SNSS-discharge as the bases for the plasmachemical reactor are formulated.

2. Physical processes underlying the SNSS microwave discharge

It is generally accepted that all known and described in literature gas discharges belong to one of the two categories: self-sustained and non-self-sustained discharges [10]. However, the research, which was carried out at the GPI [1–6], permits us to amend this supposedly unbreakable partition by introducing a third form of discharge. Namely, the SNSS-discharge, which represents unusual transition sequences from the self-sustained to non-self-sustained stage via the strong nonlinear phase of ionization-overheating instability.

The physical analysis, which was carried out in [3, 11, 12], explains the mechanisms of formation of an ionization wave in an initially extremely subthreshold microwave beam in high-pressure gases. The model, which is presented in [3, 11], is based on the assumption about a decisive role of ultraviolet radiation that arises in both the initiation stage and later stages, specifically, in its front. The model is illustrated in figure 3, which shows schematic boundaries of the microwave beam and ionization wave at some time. In addition, discharge channels (filaments) are shown, which arise near the initiator as well as in the formed propagating axial ionization wave, specifically in its front, as a result of the transformation of the non-self-sustained discharge into the self-sustained discharge via the strongly nonlinear stage of ionization-overheating instability. Discharge channels in the front of the ionization wave play a part of a UV radiation source ahead of the ionization wave front at a distance of $l \leq l_f$ (l_f is the average free path of the ionization radiation). The microwave beam interacts with this plasma, becomes partially absorbed, and heats the gaseous medium via the electron component.

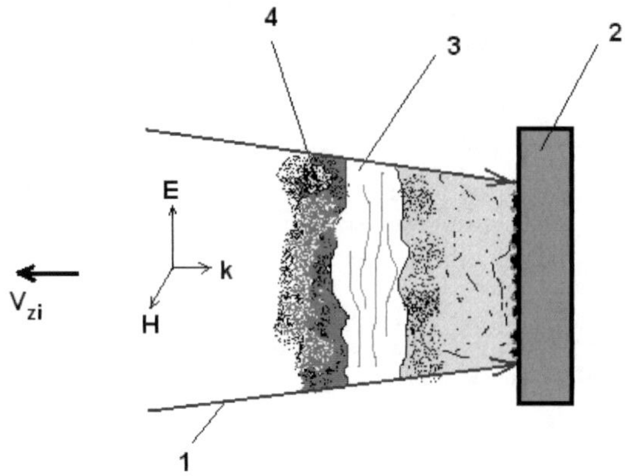

Figure 3. Scheme of the SNSS-discharge formation. 1-microwave beam; 2-metal-dielectric initiator of local discharge; 3-self-sustained-discharge; 4-non-self-sustained-discharge.

As it has been revealed in [3], the velocity of SNSS-discharge propagation toward the microwave beam in the air medium at a high pressure ($\nu_{eff} >> \omega, \nu_{eff}^*$) can be estimated using the following relation:

$$v_{Zi} \approx 10\ n_e\ E_0^2/p^4 \approx 6.8\ 10^{-5}\ n_e\ I_f^{3/2}/p_0^4 \qquad (5)$$

Here, ν_{eff}^* is the electron-neutral collision rate, which corresponds to the gas molecular density n_m^* such that $\gamma = \gamma_{thr}$, n_e is the electron density of photoionized plasma ahead of the SNSS-discharge front, p_0 is the initial pressure of gas in Torr, E_0 is expressed in kV/cm, and I_f is the microwave intensity in W/cm^2.

The nonlinear stage of the ionization-overheating instability growing in the non-self-sustained microwave discharge has been studied theoretically in [13]. In this paper, the plasma parameters of the created filaments have been estimated. The final electron temperature in the filaments is calculated using the formula

$$T_e \approx (\nu_{eff}^0/\omega)^2\ T_{e0} \qquad (6)$$

and the electron density has the form of

$$n_e \approx (\nu_{eff}^0/\omega)^{\beta i+1}\ \nu_a/\nu_T\ \ n_{e0} \qquad (7)$$

Here, $\beta_i \approx 5$–6; ν_{eff}^0 is the electron-neutral effective elastic collision rate; n_{e0}, T_{e0} are the initial density and temperature of electrons in the uniform plasma of a non-self-sustained discharge, respectively. $\nu_a = 1/\tau_a$, where τ_a is the characteristic of electron loss time (due to attachment, recombination or diffusion).

$$\nu_T \approx 10^{-9}\ \sigma_0\ E_0^2/p_0 \qquad (8)$$

here, σ_0 is the initial plasma conductivity in electrostatic units, and E_0 is in units of V/cm.

After substituting the values of ν_{eff}^0, T_{e0}, σ_0 and ν_a, which are characteristic of the experiments described in [3, 11, 14], we find that the final plasma state in the channels corresponds to an almost total ionization at the electron temperature of several hundreds of electron volts. However, the real values of the electron density and temperature can be limited by the number of unaccounted processes in [13], in particular, the reflection of microwaves from the filaments, etc. Nevertheless, the estimation results show that the discharge filaments are indeed the regions of dense and high-temperature plasma, which fit the measurements conducted in [14] that gave values of $n_e \approx 10^{16}$–10^{17} cm^{-3} and revealed the presence of a photoionized plasma 'aureole' around the discharge channels. The gas temperature in the

IOP Conf. Series: Journal of Physics: Conf. Series **907** (2017) 012022 doi:10.1088/1742-6596/907/1/012022

filaments was measured in a number of works [11] and found to be approximately $T_g* \approx 5000$ – 7000 K.

There is reason to believe that the SNSS-type discharges, which possess regions of fast and high heating of gas with a subsequent fast cooling of gas, may be a very efficient means for processing gaseous media.

3. Subthreshold self-sustained discharge (SNSS) in a microwave beam of millimeter waves in atmospheric pressure gases

3.1. *Scheme of the experiment*

The search for a possibility to realize microwave generation of plasma, which looks unusual in its physical properties and, at the same time, is promising for some technological applications, namely the SNSS-discharge in large gas volumes at pressures above or on the order of the atmospheric pressure, led us to the construction of the TORCH installation at the GPI. This installation is assembled around the MIG-3 GYROTRON complex that generates a microwave beam with the following parameters: microwave pulse power of $P_i \leq 600$kW, pulse duration of $\tau_{MW} \leq 20$ ms, wavelength of $\lambda \approx 0.4$ cm. The complex is designed to create and confine a high-temperature plasma in an L-2M stellarator [15] (see the scheme in figure 4). The installation is provided with a quasi-optical system, which is designed and manufactured at the GPI [16]. A quasi-optical system forms a microwave beam that propagates from the outlet window of the gyrotron toward the input window of the stellarator chamber. The electric field distribution in the cross section of the microwave beam is approximately Gaussian.

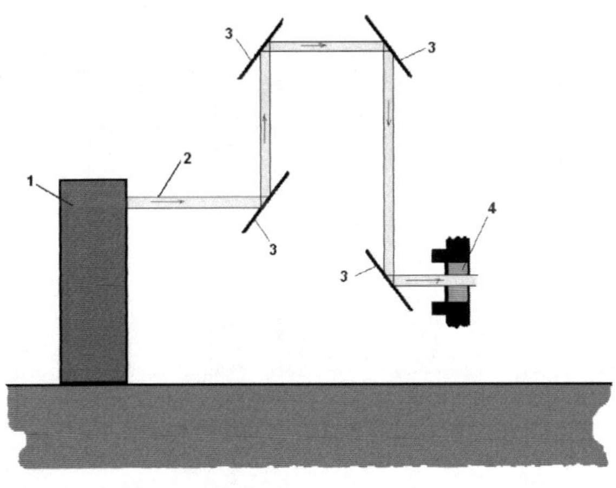

Figure 4. Scheme of the experimental setup designed for the creation and confinement of a high-temperature plasma in an L-2M stellarator. 1-MIG-3 Gyrotron complex; 2-microwave beam; 3-microwave reflecting mirrors; 4-input window of the stellarator.

In the experiments carried out in the TORCH installation, SNSS-discharges were excited in divers cross sections of microwave beam, which was generated via the MIG-3 GYROTRON complex using special elaborate initiators (see figures 5 and 6). Many experiments were performed with initiators being placed into the beam cross sections at distances of 650 mm and 440 mm from the outer window of the GYROTRON.

The minimum beam cross section at a distance of 650 mm is elliptical with semi-axes of a = 3 cm and b = 1 cm. The cross sectional area is $S = \pi ab = 9.5$ cm^2. At a distance of 440 mm, the beam cross section is circular with a radius of 3 cm: $S = \pi r^2 = 28$ cm^2. The electric field in the beam may be calculated using the formula $E_{max} = 27.5\ (P/S)^{0.5}$ V/cm, where P is the microwave power expressed in watts, and the beam area S is in cm^2. At the minimum beam cross section at the specified power of 200 kW, we have $E_{max} = 3990$ V/cm, whereas in the circular region, $E_{max} = 2300$ V/cm.

IOP Conf. Series: Journal of Physics: Conf. Series **907** (2017) 012022 doi:10.1088/1742-6596/907/1/012022

Even at the maximum powers of the microwave pulse, the values of reduced electric field in the beam do not exceed $\gamma \leq 10^{-16}$ V cm^2. Note that this value is more than one order lower than the threshold for generation of a self-sustained discharge in free space [10]: $\gamma_{thr} \approx 10^{-15}$ V cm^2.

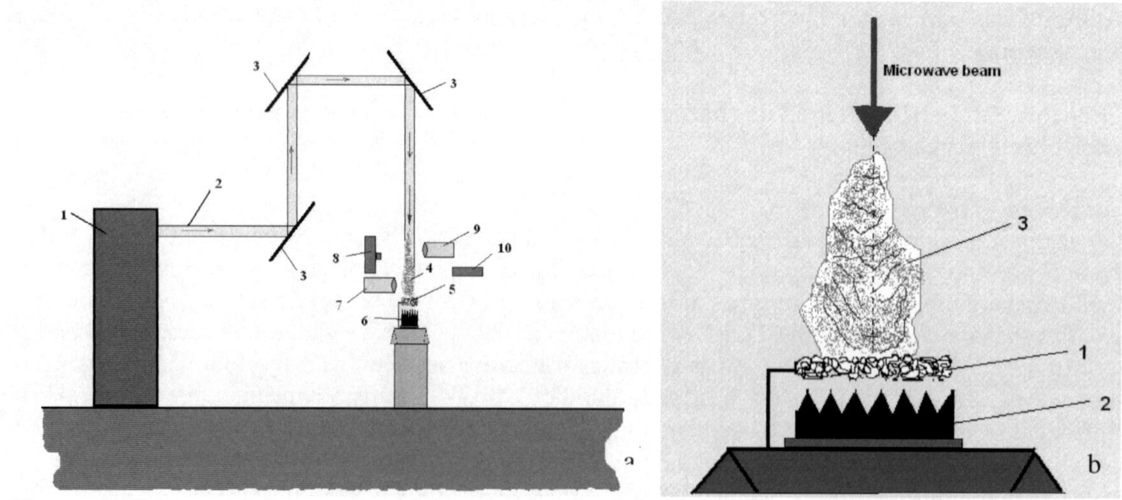

Figure 5. Scheme of the experimental setup designed for the SNSS-discharge excitation using a subthreshold microwave beam in free space of atmospheric air. a) 1-MIG-3 Gyrotron complex; 2-microwave beam; 3-microwave reflecting mirrors; 4-SNSS-discharge; 5-initiator; 6-microwave absorber;7-high-speed video camera; 8-streak camera FER-7; 9-optical spectrograph; 10-photomultiplier. b) 1-initiator; 2-microwave absorber; 3-SNSS-discharge.

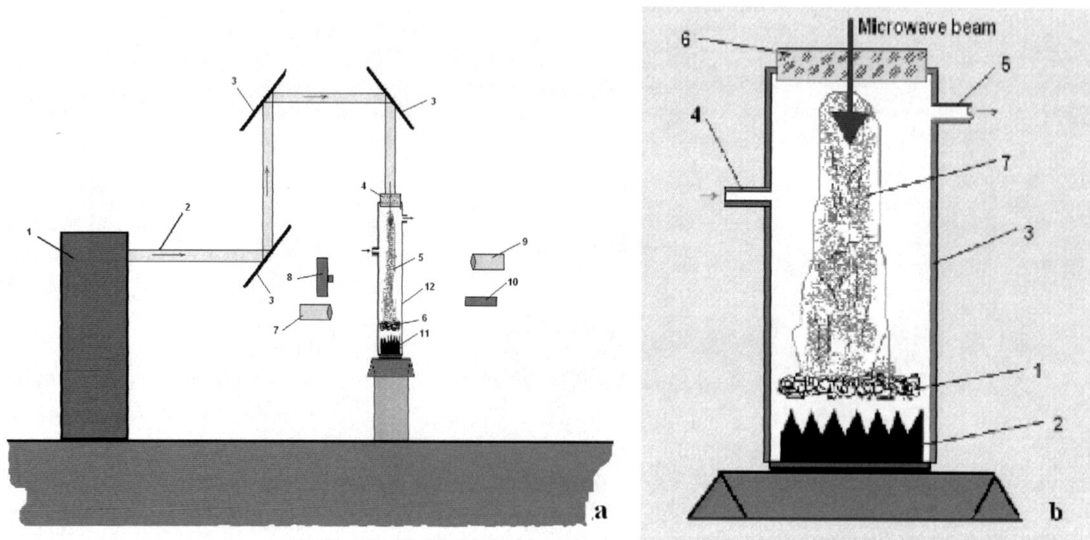

Figure 6. Scheme of the experimental setup designed for the SNSS-discharge excitation in a closed cylindrical chamber using a subthreshold microwave beam. a) 1-MIG-3 Gyrotron complex; 2-microwave beam; 3-microwave reflecting mirrors; 4) microwave transparent window; 5) SNSS-discharge; 6-initiator; 7-high-speed video camera; 8-streak camera FER-7; 9-optical spectrograph; 10-photomultiplier; 11-microwave absorber; 12-cylindrical chamber. b) 1-initiator; 2-microwave absorber; 3-cylindrical chamber; 4-working gas intake fitting; 5-working gas withdrawal fitting; 6-microwave transparent window; 7-SNSS-discharge.

The construction and properties of the initiator, which is specially designed for our experiment, are such that it does not require any external energy source and does not reflect any substantial portion of the incident microwave radiation energy, which would otherwise entail disruption of microwave generation. The initiator is a tangle of chaotically arranged thin metal filaments with a diameter of ~0.5 mm. This construction (chaotically arranged metal wires) forms an arch system, which can be easily deformed to cover a substantial part of the microwave beam cross section. Testing of this construction in air at normal pressure (1 atm) demonstrated its efficiency for creating a thin layer of self-sustained discharge plasma adjacent to the initiator. The breakdown mechanism is rationalizes as a result of formation of chaotically positioned half-wave cavities, in which the field increases up to breakdown values for atmospheric air. The Q-factor of such cavities for the field amplification must be as high as ~100.

In the TORCH installation, we studied the possibility of exciting an SNSS-discharge in free space in atmospheric air as well as in the closed chamber filled with various gases or gas mixtures at pressures close to the atmospheric pressure. The microwave subthreshold discharge experiments in atmospheric air and in a closed chamber are represented diagrammatically in figures 5 and 6, respectively. Here, behind the initiator, an absorbent load takes place. The two elements presented in figure 5b and figure 6b are indeed those parts, which ensure the SNSS-discharge generation.

The cylindrical chamber can be made of either dielectric or metal. Its diameter may not be less than the beam diameter. The end that faces microwave radiation is closed by a radio transparent window, which is meant for the microwave beam input. The initiator and the microwave absorber are placed near the opposite end of the cylinder.

We used the following diagnostic methods:

- Integral photography of the discharge during the time of the microwave pulse;
- Detection of the transmitted and reflected levels of microwave radiation;
- Recording of dynamics of axial propagation of the SNSS-discharge using a high-speed video camera;
- Streak camera registration of the SNSS-discharge axial propagation using FER-7 GPIoperating in the continuous sweep mode;
- Recording of the optical spectrum of emission from the discharge using AVASPEC 2048, 3648 and HR2000 spectrographs;
- Recording of the temporal and spatial behavior of the optical radiation from the SNSS-discharge using FEU-106 photomultipliers.
- Recording of the SNSS-discharge UV-radiation (240 – 360 nm) using GUVA-S12SD UV-photodiodes.

3.2. *Results of the experiment*

The experiments, which were conducted using the TORCH bench and are illustrated in the schemes of figures 5 and 6, demonstrated the possibility of exciting an SNSS-discharge (in free space and in a closed chamber) at atmospheric pressure in single microwave pulses beginning with a peak power on the order of $P_i \approx 200$ kW and up to 600 kW and pulse duration of $\tau_{MW} \approx 200$ μs and up to 20 ms. The discharge excited using a special-purpose initiator occupies an extended volume. When the power is large or when thc microwave pulse is long, the discharge axial length is large.

We captured photographs of the discharge, which was excited following the scheme in figure 5, using microwave beams with powers of 200 and 400 kW at different microwave pulse duration. The photographs of the discharge are presented in figures 7 and 8.

From the photographs, we determined the axial advance velocity of the SNSS-discharge that is freely localized in the atmospheric air. The measurement data are summarized in figure 9. The axial size of discharge achieved the value as high as $L \approx 50$ cm.

From the photographs, we determined the axial advance velocity of the SNSS-discharge that is freely localized in the atmospheric air. The measurement data are summarized in figure 9. The axial size of discharge achieved the value as high as $L \approx 50$ cm.

IOP Conf. Series: Journal of Physics: Conf. Series **907** (2017) 012022 doi:10.1088/1742-6596/907/1/012022

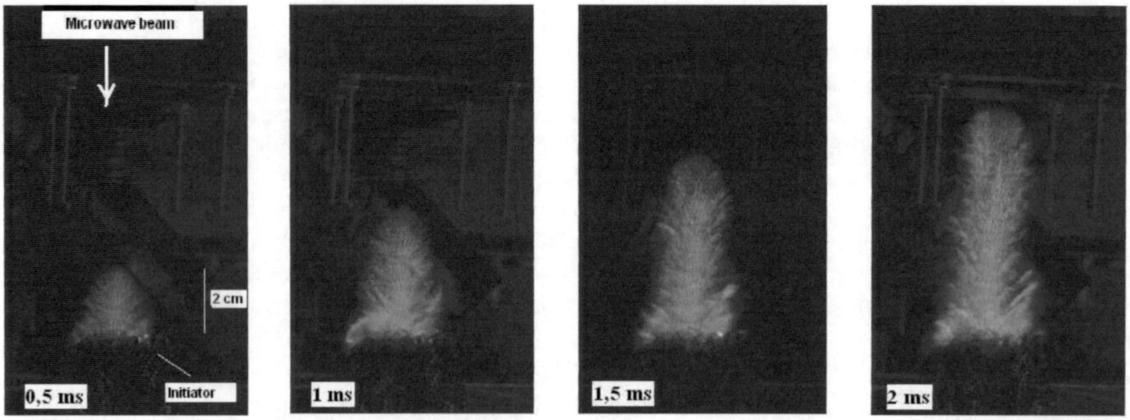

Figure 7. Photographs of SNSS-discharges in a free space of atmospheric air under different durations of microwave pulses. $P_i = 200$ kW.

Figure 8. Photographs of SNSS-discharges in a free space of atmospheric air under different durations of microwave pulses. $P_i = 400$ kW.

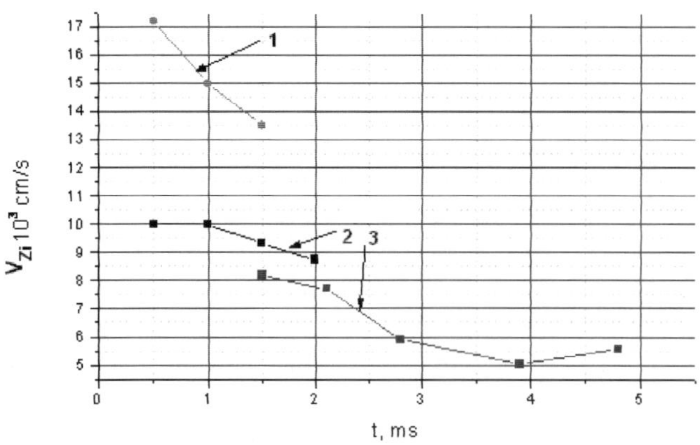

Figure 9. Time dependence of the SNSS-discharge excited in the atmospheric pressure air. The axial velocity is determined from the photographs (curves 1 and 2) and from the UV-photodiode signals (curve 3). $P_i = 400$ kW (curve 1) and 200 kW (curves 2 and 3, respectively). $\tau_{MW} = 6$ ms.

IOP Conf. Series: Journal of Physics: Conf. Series **907** (2017) 012022 doi:10.1088/1742-6596/907/1/012022

UV radiation of the SNSS-discharge excited in the atmospheric pressure air was registered during one microwave pulse with a help of five GUVA-S12SD UV-photodiodes that were placed 5 cm apart on a slat. The spatial resolution of radiation on the beam axis is $\Delta L \approx 2$ cm. Time dependence of the SNSS-discharge axial velocity is shown in figure 9 (curve 3).

To determine the axial advance velocity of the SNSS-discharge during one microwave pulse, we treated the streak camera images of the SNSS-discharge obtained using FER-7 operating in a continuous sweep mode. This photograph is presented in figure 10. The slit of the camera was oriented along the discharge axis (which is vertical in the figure), where the horizontal coordinate is the time. Typical temporal and spatial scales are indicated in the figure. As evident from figure 10, the discharge moves toward the microwave radiation with a velocity of ~ 10^4 cm/s. This result agrees well with the results obtained from the photos and UV-photodiode measurements.

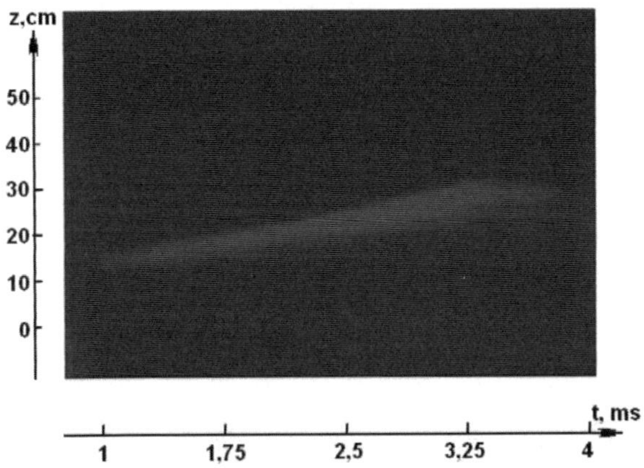

Figure 10. Streak camera image of the SNSS-discharge in the atmospheric pressure air obtained using FER-7 operating in a continuous sweep mode. Free space conditions. $P_i = 200$ kW.

Note that the maximum brightness is observed at the background of the discharge front, whereas the brightness weakens behind the front.

We captured photographs in air using a high-speed video camera in the frame mode. The characteristic photographs are presented in figure11. The time between frames is 4 ms; the exposure time is 2/5000 s = 200 μs. The microwave power is 200–300 kW, and the duration is 6 ms.

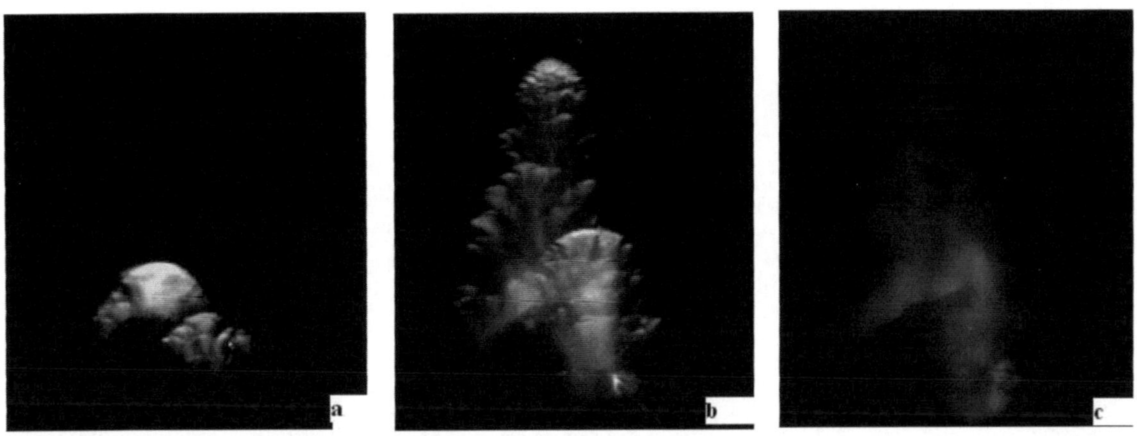

Figure 11. High-speed photography of the SNSS-discharge (air, free space conditions). $\tau_{MW} = 6$ ms; $P_i = 200$ kW. a) Starting stage of the discharge; b) a 4-ms delay after the first frame; c) post-discharge stage.

One can see from the photographs and UV-photodiode oscillograms that the intensity is maximal at the front. However, it weakens behind the front, which may be explained by the reduction in microwaves. A similar picture is observed in oscillograms, which were recorded using a photomultiplier. The discharge emission dynamics, which were recorded using FEU-106 positioned at a distance of 8 cm from the initiator, are presented in figure 12. Collimation ensures that the spatial resolution is approximately 2.5–3.0 cm. The microwave power is 200 kW, the duration is 4 or 6 ms. One can see how the intensity initially rapidly increases and then decreases behind the discharge front.

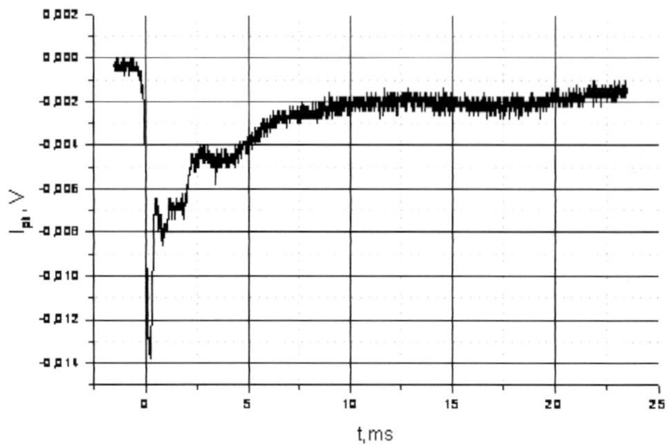

Figure 12. Photomultiplier signal oscillogram. SNSS-discharge in air (free space conditions). L = 8 cm; $P_i = 200$ kW; τ_{MW}=4ms.

The optical emission spectrum of the discharge in air was studied using AVASPEC 2048, 3648 and HR2000 spectrographs. The most representative spectrum was obtained using an HR2000 spectrometer in the range of 300–400 nm. The second positive system of nitrogen falls precisely in this range. Figure 13 shows the spectrum of microwave discharge in air with identification of bands of the second positive system of nitrogen. The treatment of spectra similar to those presented in figure 13 confirms that the gas temperature in the observed region is ~4000–4800 K.

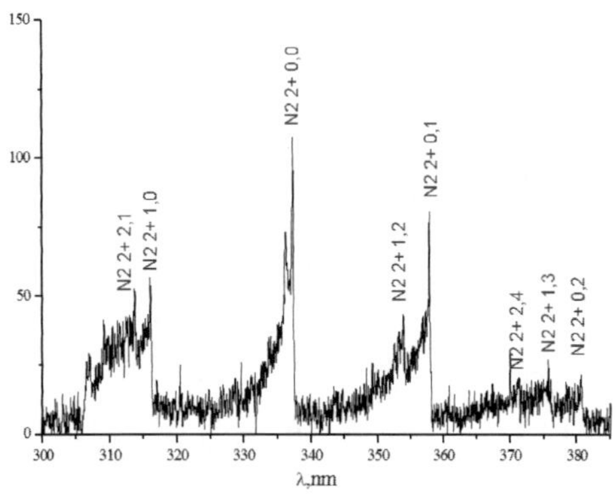

Figure 13. Optical spectrum of the SNSS-discharge in air (free space conditions). $P_i = 200$ kW; $\tau_{MW} = 6$ ms.

Figure 14 presents a photograph of the SNSS-discharge, which is excited according to the experimental scheme in figure 6, in the CH_4:CO_2 mixture filling a closed chamber at atmospheric pressure. When we used methane-hydrogen and methane-water mixtures, the H_α. line was present in the emission spectra (see figure 15). Based on the assumption that the Stark mechanism is responsible for the broadening of this line, we conclude that the electron density in the discharge exceeds 10^{16} cm^{-3}.

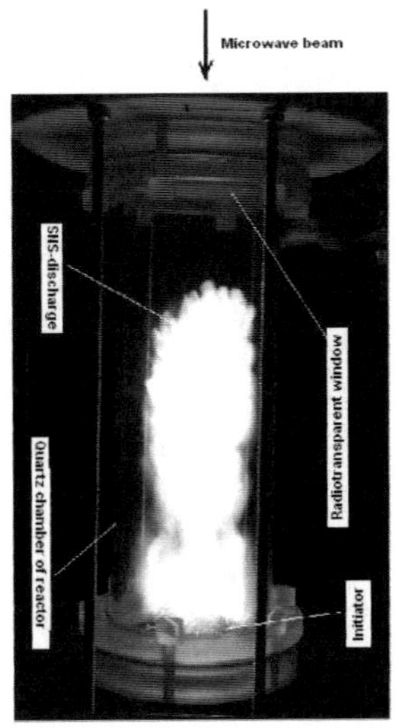

Figure 14. Photograph of the SNSS-discharge excited in the closed quartz chamber filled with a CH_4:CO_2 gas mixture. P_i = 400 kW; τ_{MW} = 4 ms; p = 1 atm.

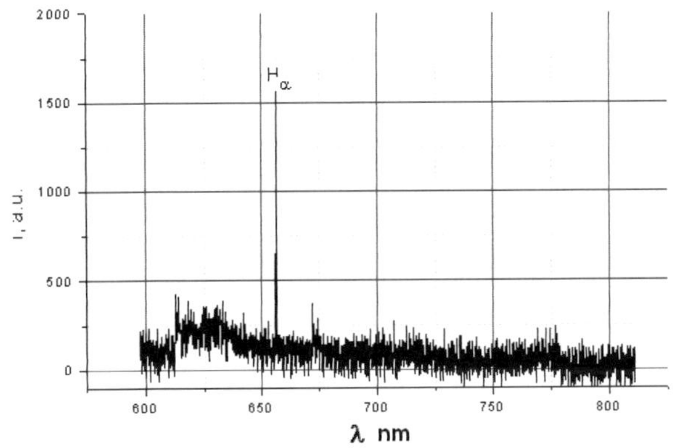

Figure 15. Optical spectrum of the SNSS-discharge in a CH_4:H_2O mixture (closed cylindrical chamber). P_i = 200 kW; τ_{MW} = 6 ms; p = 1 atm.

3.3. *Discussion of the experimental results*

The structure of the discharge, which is excited according to the schemes in figures 5 and 6, as well as the dynamics of its propagation in the microwave beam correspond to the characteristics of SNSS-discharges described in [3, 5, 14]. The fact that the axial velocities observed in the present work fall outside the previous values (e.g., in [3]) may be primarily due to the experimental conditions. When using MIG-3, we managed, for the first, time to increase the level of the working gas pressure to atmospheric. Taking into account the strong dependence of velocity, V_{Zi}, on pressure ($\sim 1/p_0^4$), as predicted by expression (5), the reduction in axial velocity in the described experiments is well substantiated. The relatively large values of ν_{eff}/ω, exceeding 10, are in line with (7) and explain the high electron density of $n_e \approx 10^{16}$ cm^{-3}, which was obtained from the Stark broadening of the H_α line. The gas temperature in the discharge region, which was determined in our experiment, was found to be 4 000–4 800 K. These values are lower than the values of 5000–7000 K, which were found in the early experiments [3, 5, 14] in the centimeter wavelength range at gas pressures substantially below the atmospheric pressure.

Using the physical model of the SNSS-discharge, which was developed in [3, 13], estimations of the specific energy input, ε, in the elementary discharge "thread" can be made:

$$\varepsilon \approx \sigma E_0^2 \tau_i \qquad (9)$$

where, σ is the conductivity of the structural "thread", and τ_i is the time of the "thread" interaction with microwave radiation during the ionization wave propagation.

Proceeding from the position that [10]:

$$\sigma \approx n_e\, e^2\, \nu_{eff} / m(\omega^2 + veff^2) \tag{10}$$

with the assumption that $\nu_{eff} >> \omega$, we obtain:

$$\sigma \approx 2{,}82\ 10^{-4}\ n_e/\nu_{eff} \quad \Omega^{-1}\ cm^{-1} \tag{11}$$

Here, n_e is in (cm^{-3}), and ν_{eff} is in (s^{-1}).

Thus, the estimation formula for the specific energy input takes the form of:

$$\varepsilon \approx 2{,}82\ 10^{-4}\ n_e/\nu_{eff}\ E_0^2\ \tau_i \tag{12}$$

where, E_0 is in (V/cm), and τ_i is in (s).

If we insert in (12) the characteristics of our experiment data ($n_e \approx 10^{16}$ cm^{-3}, $\nu_{eff} \approx 2\ 10^{11}$ s^{-1}, $\tau_i \approx 5$ μs and $E_0 \approx 2 \cdot 10^3$ V/cm), the specific energy input in the "thread" will be as high as

$$\varepsilon \approx 0{,}4\ kJ/cm^3$$

Hence, the discharge observed in our experiment can be classified as an SNSS-discharge for the following reasons:

- The discharge is excited under conditions that the reduced electric field in the gas volume occupied by the beam is substantially below the threshold for the excitation and maintenance of the self-sustained discharge;
- Structurally, the discharge forms from thin separated channels (filaments) that are stretched along the microwave electric field;
- The axial propagation velocity of the plasma in each channel, and both gas and gas-discharge plasma parameters agree with the SNSS-discharge physical model.
- As it has been predicted by the physical model [3, 5], the discharge excited in this experiment is the source of intense UV-radiation.

The results obtained in the described experiments allow us to begin the development of a laboratory plasmachemical reactor, which is based on a high-power beam of millimeter-wavelength radiation, by following the scheme depicted in figure 6 based on the MIG -3 GYROTRON complex.

The main studies in the area of plasmachemistry that are promising for the current technology would be pertinent to the following problems:

- carbon dioxide conversion of methane into synthesis gas;
- water vapor conversion of methane into synthesis gas;
- utilization of carbon dioxide.

A series of foregoing experiments and theoretical (calculation) work (see [17–20]) have shown that the peculiarity of this type of discharges is the "filament" plasma structure, which is characterized by a high electron density, both high electron and gas temperature, and by record-breaking high specific energy deposition. It is very attractive for solving some plasmachemical problems, including the abovementioned problems. This phenomenon is explained by the fact that the extremely high gas temperatures attainable in filaments at the nonlinear stage of the ionization-overheating instability lead to the almost total decomposition of the working gas molecules into radicals and atoms. The fast cooling of the "thread" structures is accompanied by quenching the gas volume processed with the discharge composition. In the framework of the "Chemical Bench" program [21], calculations were performed, in which a filament was modeled sequentially using two reactors at thermal equilibrium. In one of the reactors, the gas was heated up to the temperatures corresponding to those attained within our experimental times and measured in our experiment. For the other reactor, the process of cooling was reproduced according to the dynamics observed in our experiment. The design-theoretical efficiency of plasmachemical processes was close to the experimentally measured value.

The approximate expression for the axial dimension L_b of the SNSS, in accordance with (5), may be represented in the form:

$$L_b \approx 6.8\ 10^{-5} \tau_{MW}\ n_e\ I_f^{3/2}/p_0^{4} \quad (13)$$

Based on relation (13) and taking it into account in our experiments in air for τ_{MW} = 2 ms and P_i = 200 kW, the length of $L_b \approx$ 15 cm was observed. We considered the possibility of the formation (in both free space and cylindrical cambers) of extended SNSS-discharges. We assumed the parameters of a single pulse: τ_{MW} = 20 ms and P_i = 600 kW, which is maximal during the operation using the MIG-3 GYROTRON complex. In addition, we investigated a possibility of extending the discharge length up to 10–15 m, which is required for the practical implementation of the method.

The possibility of exciting long SNSS-discharges in atmospheric pressure gases can significantly expand their application area by including the problems of global gas-discharge cleaning of atmosphere from ozone-destructing impurities as well as industrial waste gas cleaning from ecologically harmful contaminations.

The physical aspects of the specific problems of cleaning atmosphere from chlorofluorocarbons (Freon) have been discussed in a series of works at the General Physics Institute and Institute of Applied Physics [17, 22–24]. As an approach for destroying chlorofluorocarbons, subthreshold microwave beams, which are generated by an earth-based station and focused at the chosen altitudes, have been considered. Realization of these objectives for the SNSS-discharges, which are excited using the presented scheme in figure 16, can be performed due to the characteristics that are typical for this discharge form. Among them, the most critical are the following:

- Use of a relatively low microwave power value (lower than the threshold value), which leads to the cheapening of the cleaning technology;
- Higher specific energy input in a gas medium than in the threshold beams due to the low (subsonic) velocity of SNSS ionization wave extension;
- Simplicity of SNSS-discharge initiation and its size control;
- Possibility within the limits of modern microwave technique to excite a plasma column as long as 20–30 m.

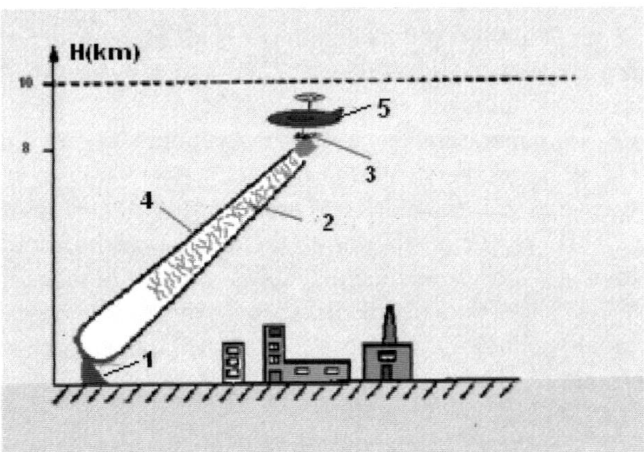

Figure 16. Schematic representation of the atmosphere cleaning system, which is based on the SNSS-discharge.1-GIROTRON and microwave antenna; 2-SNSS-discharge; 3-initiator; 4-microwave beam; 5-pilotless vehicle.

The achievement of meter axial lengths of SNSS-discharges in microwave beams may be also used to design an electric discharge system for cleaning industrial gas ejections (a schematic version is depicted in figure 17).

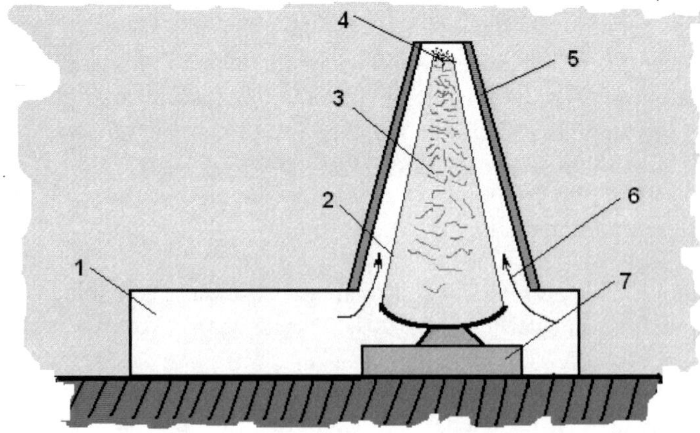

Figure 17. Schematic representation of industrial gas ejection cleaning, which is based on the SNSS-discharge. 1-plant building; 2-microwave beam; 3-SNSS-discharge; 4-initiator; 5-plant chimney; 7-GIROTRON and antenna.

4. Conclusions

Due to the performed at the GPI experiments, the area of realization of a new form of the deeply subthreshold self/non-self-sustained (SNSS) discharge has been considerably expanded. For the first time, the excitation of this discharge at gas pressure of the order of (and higher) than atmospheric pressure was demonstrated in free space of atmospheric air as well as in a closed cylindrical chamber filled with different gases.

For the first time, the SNSS-discharge was created using millimeter wavelength microwave beams at a high pulse power of ≈600 kW. For the first time, the discharge was initiated in any beam cross section using a specially made construction, which is composed of metallic thin threads. In this case, discharge initiation as well as its maintenance in the form of axially propagating ionization wave was carried out using only one microwave energy.

Discharge parameters are compatible with those predicted by the physical model, which considered the SNSS-discharge as a progression of sequential transitions of non-self-sustained and self-sustained discharges through the excitation and development of ionization-overheating instability.

GYROTRON was applied, which is capable of generating in a pulse-periodical regime a microwave radiation with an average power of several hundred kW (at the peak pulse power of ≈600 kW). Due to its relative simplicity, this microwave radiation, interfaced with a cylindrical reactor chamber as well as high efficiency value of microwave energy contribution to the deeply subthreshold discharge and through it in a gas medium, allows to draw conclusions about the feasibility of plasmachemical reactor creation based on the GYROTRON and SNSS-discharge, which can satisfy modern industrial requirements of various chemical technologies such as methane conversion in syngas, CO_2 utilization, industrial gaseous waste cleaning, etc.

Acknowledgments

The work is supported in part by the Program of Basic Research of the Russian Academy of Sciences, 2017, III.2, Fundamental research Program of RAS "Physics of atmosphere: electric processes, radio-physical methods of research", Fundamental research Program of RAS, 2017, 131П "Combustion and Explosion" and the Russian Science Foundation (Project 17-12-01352)

References

[1] Kossyi I A, Gritsinin S I, and Davydov A M 2012 *8th Int. Workshop Microwave Discharges: Fundamentals and Applications Book of Abstracts* Zvenigorod, Russia, (STC "PLAZMAIOFAN") 39

[2] Kossyi I A 2011 *Proc. of 8th Int. Workshop "Strong Microwaves and Terahertz Waves Sources and Applications* N. Novgorod – St. Petersburg, Russia (Institute of Applied Physics RAS) 25

[3] Batanov G V, Gritsinin s i, Kossyi I A, Magunov A N, Silakov V P and Tarasova N M 1985 *Plasma Physics and Plasma Electronics* Ed. L M Kovrizhnykh (Nova Science Publishers, Commack) 241–282

[4] Batanov G M, Gritsinin S I, and Kossyi I A 2002 *J. Phys. D: Appl. Phys.* **35** 2687–2692

[5] Kossyi I A 2006 *44th AIAA Aerospace Sciences Meeting and Exhibit* Reno Nevada Report AIAA-1457

[6] Kossyi I, A, Batanov G M, Berezhetskaya N K, Davydov A M, Sarksjan K A, and Harchev N K 2016 *Physical Basis and Plasma – Chemical Application HAKONE 15 15th High pressure low temperature plasma Chemistry Simposium* Brno Czech Republic Book 0f Abstract (Masaryk University)

[7] Brovkin V G , Bykov D F, Golubev S K et al 1991 *Zhurnal Tekhnicheskoi Fiziki* **61** 153–157 (in Russian)

[8] Barkhudarov E M, Berezhetskaya N K, Bol'shakov E F, Eletskii A V, Kossyi I A and Taktakishvili M I 1984 *Zhurnal Tekhnicheskoi Fiziki* **54** 1219–1222 (in Russian)

[9] Gritsinin S I, Kossyi I A, Silakov V P, Tarasova N M and Terekhin V E 1986 *Preprint of IOFAN* № 14

[10] Raizer Yu P *Gas Discharge Physics* (Nauka: Moscow, 1987, Springer-Verlag: Berlin, 1991)

[11] Kossyi I A, Batanov G M, Berezhetskaya N K, Davydov A M, Sarksjan K A, and Harchev N K 2016 *43rd Conference on Plasma Physics* EPS-2016 Proc Leuven, Belgium. (EPS) 1–4

[12] Artemov V I , Levitin and Yu S, Sinkevich O A 1994 Instability and turbulence in the low temperature plasma (Moscow: Moscow Energetic Institute) 402 p (in Russian)

[13] Kim A V and Fraiman G M 1983 *Fizika Plazmy* **9** 613–617 (in Russian)

[14] Avetisov V G, Gritsinin S I, Kim A V et al 1990 *Pis'ma v Zhurnal Eksperimental'noi i Tekhnicheskoi Fiziki* **51** 306–309 (in Russian)

[15] Batanov G M, Belousov V I, Bondar' Yu F, Borzosekov V D et al 2012 *Prikladnaya Fizika (Applied Physics)* **6** 79–83

[16] Batanov G M, Kolik L V, Kurbatov V I, Petrov A E, Sapozhnikov A V, Sarksyan K A and Shatz M G 1986 L-2 *Fizika Plazmy* **12** 762–764 (in Russian)

[17] Askar'yan G A, Batanov G M, Barkhudarov A E, Gritsinin S I, Korchagina E G, Kossyi I A, Silakov V P and Tarasova N M 1992 *Fizika Plazmy* **18** 1198–1211 (in Russian)

[18] Gritsinin S I, Knyazev V Yu., Kossyi I A and Popov N A 2006 *Plasma Physics Reports* **32** 520–524

[19] Gritsinin S I, Gushchin P A, Davydov A M., Ivanov E V., Kossyi I A. and Misakyan M A 2009 *Plasma Physics Reports* **35** 933–940.

[20] Parkhomenko Yu N, Polisan A A, Skryleva E A, Tabachkova N Yu, Shul'ga N Yu, Davydov A M, Kossyi I A, Duyzhikov I N and Pokalyakin V T 2012 *Russian Microelectronics* **41** 491–497

[21] Potapkin B Vet al 1998–2000 Program "Chemical Work Bench (Kinetic Technologies (KINTECH))"

[22] Batanov G M, Kossyi I A and Silakov V P 2002 (In memory of G.A. Askar'yan) *Plasma Physics Reports* **28** (Pleiades Publishing) 204–228

[23] Askaryan G A, Batanov G M, Kossyi I A and Kostinskii A Yu 1992 ISPP-10 (Piero Caldirola) *High Power Microwave Generation and Applications* Eds E.Sindoni and C.Wharton SIF Bologna (Editrice Compositori) 207–226

[24] Askaryan G A, Batanov G M,.Bykov D F, Gritsinin S I, Kossyi I A, Kostinskii A Yu, Matveyev A A and Silakov V P. 1991 ISPP-7 "Piero Caldirola", *Controlled Active Global Experiments (CAGE)*. Eds E.Sindoni and A.Y.Wong SIF Bologna (Editrice Compositori) 239–258

Microplasma discharges exited by a plasma flow on constructional metals

V A Ivanov[1,2], A S Sakharov[1], M E Konyzhev[1], T I Kamolova[1], A A Dorofeyuk[1] and L I Kuksenova[3]

[1] Prokhorov General Physics Institute, Russian Academy of Sciences, Moscow, 119991 Russia
[2] National Research Nuclear University "MEPhI," Moscow, 115409 Russia
[3] Mechanical Engineering Research Institute, Russian Academy of Sciences, Moscow, 101990 Russia

Abstract. Results are presented from experimental and theoretical studies of microplasma discharges excited by a plasma flow on constructional metals partially covered with a dielectric (oxide) film. Tribological tests of metal samples treated by microplasma discharges show a significant increase in their wear resistance due to the formation of a strong microrelief on the sample surface.

1. Introduction

The interaction of a pulsed plasma flow with a negatively biased (~100–450 V) metal surface partially covered with a ~1-μm-thick dielectric film results in the excitation of microplasma discharges at the film edge [1–3]. The mechanism of excitation is as follows. Under the action of the ion flow, the outer surface of the film acquires a positive electric potential with respect to the metal, due to which a strong electric field of about several MV/cm arises at the film edge. Being additionally amplified at the microroughness of the metal surface, this field initiates explosive emission from the metal and results in the development of microarcs (microplasma discharges) with a plasma density of $\sim 10^{20}$ cm^{-3} and temperature of ~0.5–1 eV [4]. The high pressure ($\sim 10^{7}$ Pa) of the plasma and metal vapor heated to ~5000 K above the melted metal surface at the locations of microplasma discharges leads to the formation of microcraters on the metal surface. Since microplasma discharges evaporate the dielectric film, the area covered with the film gradually shrinks and, after several plasma pulses, the entire metal surface turns out to be remelted and cleaned of the film. As a result, a strong microrelief in the form of overlapped craters is produced on the metal surface.

2. Experiment

In our experiments, microplasma discharges on metal samples were excited by a pulsed plasma flow with a density of $\sim 10^{12}$–10^{13} cm^{-3} and an electron temperature of $T_e \approx 10$ eV. The plasma ion component consisted mainly of hydrogen ions (90–80%) and singly charged carbon ions (10–20%). The parallelepiped samples with a total surface area of 2–3 cm^2 were made of constructional metals, such as VT1 titanium, V95 aluminum alloy, and Steel 45. Dielectric (oxide) films of thickness ≤2 μm were delosited on the sample surface by heating in open air [1, 2]. Thicker dielectric (SiO_2) films on titanium samples were deposited by sputtering in high vacuum. Under the action of an ~20-μs-long plasma flow, microplasma discharges were excited at the border of the sample surface covered with the dielectric film (figure 1). The microplasma discharges continued to burn for ~20 ms (while the negative voltage was being applied to the sample), i.e., long after the plasma pulse has terminated. After ~10 plasma pulses, the sample was completely cleaned of the film and its entire surface acquired a microrelief (figure 2). At MPD currents of 100–600 A, the microrelief height on all six faces of the sample was in the range of 2–18 μm. For a voltage applied to the sample of $\phi_0 = -450$ V, the probability of microplasma discharge excitation on a sample covered with a 1-μm-thick film was close to 100%. When using films with thichnesses of 4 μm and more (up to 12 μm), the probability of microplasma discharge excitation decreased with increasing film thickness (figure 3). When using films with thicknesses less than 1 μm, the probability of microplasma discharge excitation decreased from a value close to 100% for film thicknesses of 0.2–1 μm to lower than 0.3% for films thinner than 0.01 μm.

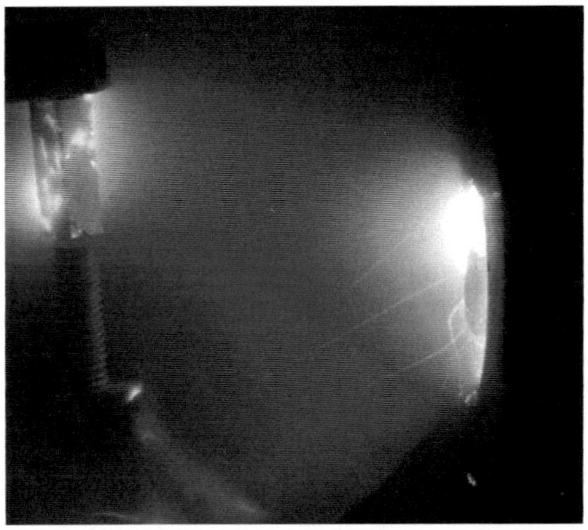

Figure 1. Plasma injector (on the right) and microplasma discharges burning at the boundary between of the dielectric film and the open Ti surface (on the left).

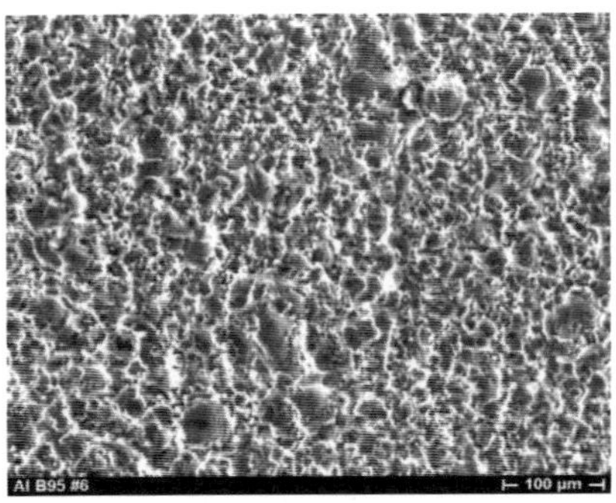

Fig. 2. Microrelief formed on the surface of a V95 Al sample under the action of microplasma discharges with a total current of 200 A and duration of 20 ms.

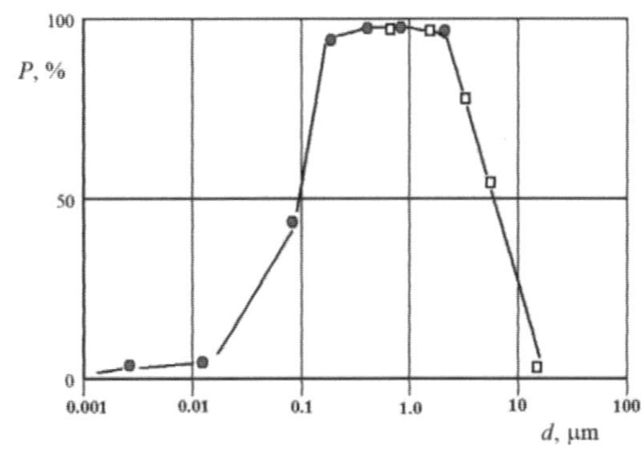

Fig. 3. Probability of microplasma discharge excitation on Ti samples covered with dielectric films of different thickness d ($\phi_0 = -450$ V).

3. Simulations

A decisive factor governing plasma–solid interaction is the strength of the electric field arising in the Debye sheath with a thickness on the order of $r_D = (T_e/4\pi e^2 n_e)^{1/2}$, where T_e is the plasma electron temperature, n_e is the electron density, and e is the electron charge. Under our experimental conditions

($n_e = 10^{12}$–10^{13} cm^{-3}, $T_e \approx 10$ eV), the Debye radius is $r_D \sim 10$ μm and the characteristic electric field in the sheath (for $\phi_0 = 0$) is as low as $E^* \sim 10$ kV/cm. Application of a high (compared to T_e/e) negative (with respect to the plasma) electric potential ϕ_0 to the metal does not alter the situation qualitatively, because the thickness of the Debye sheath increases with increasing applied voltage as $|\phi_0|^{3/4}$ and, accordingly, the electric field strength on the metal grows very slowly, $E \sim |\phi_0|^{1/4}$. For $T_e = 10$ eV and $\phi_0 = -400$ V, the electric field on the metal surface increases compared to E^* by only a factor of 4, i.e., to ~40 kV/cm. The situation, however, changes radically if the metal surface is covered with a thin ($d \ll r_D$) dielectric film. In this case, the flux of positive ions from the plasma charges the outer surface of the film and its potential tends toward the floating potential $\phi_f \approx -3T_e/e \approx -30$ V with respect to the plasma. In our experiments, $|\phi_f| \ll |\phi_0|$; hence, almost the entire voltage between the plasma and metal turns out to be applied to the thin dielectric film and the electric field inside the film is $E \sim |\phi_0|/d$, i.e., for $\phi_0 = -400$ V and $d \sim 1$ μm, it can reach several MV/cm. At the film edge (as well as in the film breaks), this field turns out to be applied to the open metal surface. In this case, even a moderate (at a level of 10–20) additional field amplification on the metal microroughnesses is sufficient to cause explosive emission from the metal surface and the formation of a microarc, i.e., excitation of a microplasma discharge at the film edge.

Generation of a strong electric field at the film edge was simulated using a modified particle-in-cell method, which takes into account that the time of flight of ions through the Debye sheath is several orders of magnitude shorter than both the duration of the plasma pulse and the characteristic time of the film charging. In this case, in the course of film charging, ions move in a quasi-stationary electric field, i.e., at each time step, ion motion can be considered established and the ion flow can be represented as an ensemble of positively charged "jets" propagating in a stationary electrostatic field. Such an approach can be called the "jet-in-cell" method [2].

To model electron emission from the metal surface near the film edge, we used the Fowler–Nordheim formula for the density of field emission [5, 6], assuming that the electron work function is $\varphi = 3$ eV and the characteristic field amplification on the microroughnesses of the metal surface is $\beta = 15$. According to [6], the threshold density of the electron emission current required for the development of explosive emission was assumed to be ~10^8 A/cm^2.

The simulations have shown that, without electron emission from the metal, the maximum value of the electric field at the open metal surface near the film edge is $E_{max} \sim |\phi_0|/2d$ [1–3] in a wide range of experimental parameters (such as the plasma density, bias voltage ϕ_0, film thickness, film permittivity, and inclination angle of the film edge). For example, for a film thickness of ~0.5 μm and $\phi_0 = -400$ V, we have $E_{max} \sim 4$ MV/cm. Taking into account the 15-fold additional field amplification on metal microroughnesses, such a field strength is quite sufficient for the development of explosive electron emission [6], followed by the formation of a microplasma discharge at the film edge.

The simulations also have shown that, depending on the inclination angle of the film edge, there are two regimes of microplasma generation. At inclination angles of $\alpha \leq 80°$, the emitted electrons do not fall on the film and practically have no effect on the value of the generated electric field (figure 4).

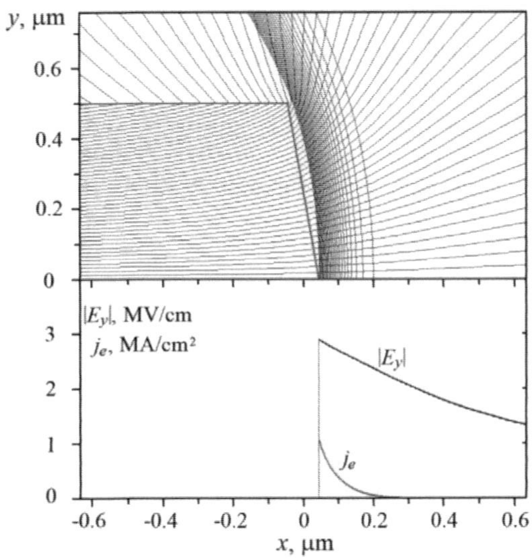

Figure 4. Distribution of the electric potential near the film edge, trajectories of emitted electrons, and profiles of the electric field strength on the metal $|E_y|$ and the field-emission current density j_e at $\phi_0 = -400$ V, $n_e = 4 \times 10^{12}$ cm^{-3}, $T_e = 10$ eV (hydrogen plasma), $d = 0.5$ μm, $\alpha = 80°$, $\varphi = 3$ eV, $\beta = 15$. The potential difference between neighboring equipotential lines is $\Delta\phi = 0.5T_e/e$.

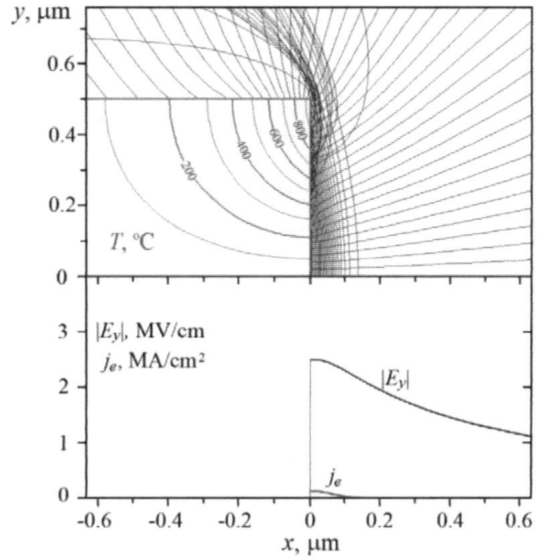

Figure 5. Distributions of the electric potential near the film edge and temperature inside the dielectric, trajectories of emitted electrons, and profiles of the electric field strength on the metal $|E_y|$ and the field-emission current density j_e at the peak of the emission current ($t = 0.71$ μs) for the same parameters as in figure 4, but for the vertical film edge ($\alpha = 90°$).

At $\alpha \approx 90°$, the electron beam emitted from the metal surface falls on the end face of the film, which results in the accumulation of a negative charge at the film edge and a substantial reduction in the field strength in this region. Taking into account secondary electron emission from the film surface partially compensates for this reduction; however, the electric field strength turns out to be insufficient for the development of explosive electron emission from the metal surface. Nevertheless, under the action of the pulsed ($\tau \sim 0.1$ μs) beam of accelerated field-emission electrons with an energy of ~100 eV and a current density of ~10^5 MA/cm^2, the end face of the film is heated to a temperature of ~1000°C (figure 5), which may result in efficient gas desorption from the dielectric surface. The development of a microplasma discharge in this case can be related to the formation of a dense plasma bunch near the film edge due to the electric breakdown of the desorbed gas and/or its ionization by the field-emission electron beam.

The shape of the experimental dependence of the probability of microplasma discharge excitation P on the film thickness (figure 3) can be explained as follows. The decrease in P at $d > 2$ μm can be attributed to the decrease in the maximum value of the electric field strength at the film edge, which is inversely proportion film thickness. At film thicknesses less than 0.1 μm, leakage of the electric charge from the outer film surface due to the finite conductivity of the TiO_2 film ($\rho \sim 0.5 \times 10^7$ Ω cm, see, e.g., [7]) comes into play, which also leads to a decrease in the probability of microplasma discharge excitation [3].

4. Applications

The treated and untreated samples were subjected to standard tribological tests [8, 9]. It was found that the wear resistance depends on the current of microplasma discharges i, which determines the parameters of the microrelief formed on the sample (the characteristic scale of microroughnesses and the thickness and structure of the remelted surface layer) in the course of microplasma treatment.

The results of tribological tests of steel 45 samples show that the allowable pressure p_a (the maximum pressure applied to the samples at which no scuffing occurs on their friction surfaces) increases from 3 MPa (for polished samples without microplasma treatment) to ~25 MPa (for samples treated by microplasma discharges) (see figure 6). The maximum allowable pressure for steel 45 samples was achieved at $i \approx 400$ A (figure 7).

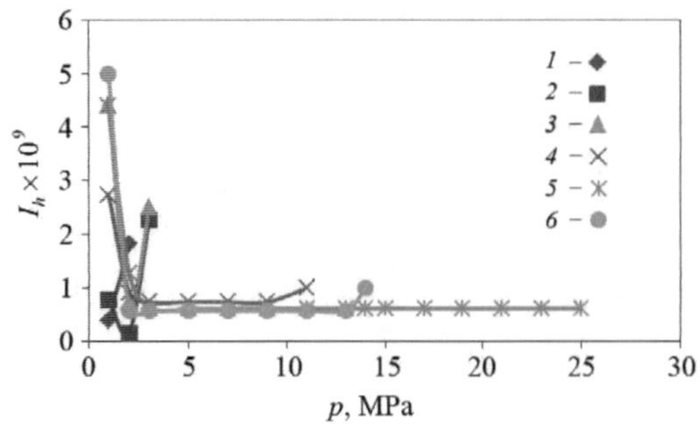

Figure 6. Wear rate I_h of steel 45 vs. pressure p applied to the samples: (*1*) without treatment; (*2*) after thermal annealing; after microplasma treatment at currents of i = (*3*) 100, (*4*) 200, (*5*) 400 and (*6*) 650 A.

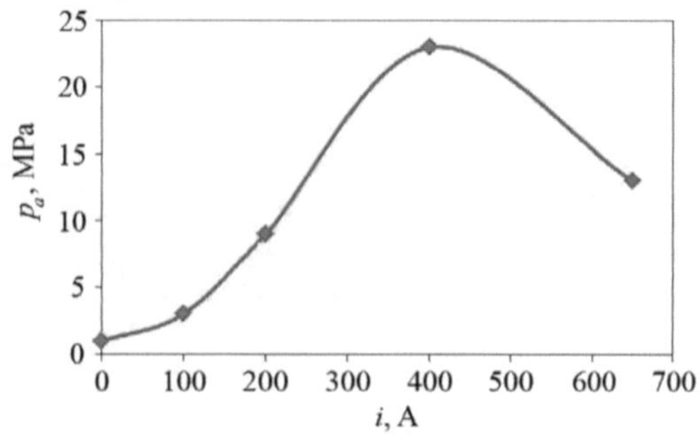

Figure 7. Allowable pressure p_a for steel 45 samples vs. current of microplasma discharges i. The leftmost point corresponds to untreated samples.

Figires 8 and 9 present results obtained from similar tests performed for samples made of D16T aluminum. Here, the allowable pressure applied to the samples increased from ~0.2 MPa for untreated samples to about 7 Ma for samples treated by microplasma discharges at a current of 100 A.

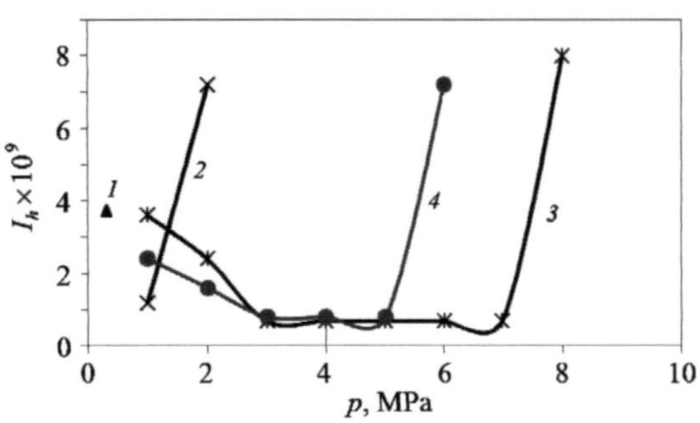

Figure 8. Wear rate I_h of D16T aluminum vs. pressure p applied to the samples: (*1*) without treatment; (*2*) after thermal annealing; after thermal annealing and microplasma treatment at currents of i = (*3*) 100 and (*4*) 200 A.

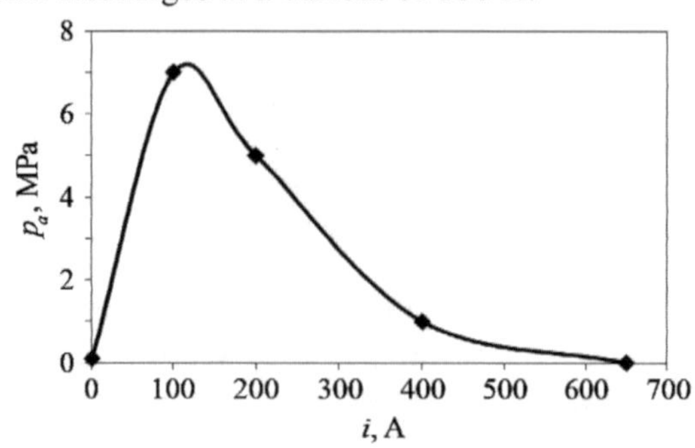

Figure 9. Allowable pressure p_a for D16T aluminum samples vs. current of microplasma discharges i. The leftmost point corresponds to untreated samples.

Finally, figure 10 presents results obtained for Ti samples treated by microplasma discharges at a current of ~300 A. The allowable pressure p_a was found increase from ~0.2 MPa for untreated samples to about 10 MPa after microplasma treatment. In this case, the wear rate I_h of treated samples in the pressure range of p = 2–10 MA turned out to decrease drastically (by many times) compared to that for untreated samples at p ~ 0.2 MPa.

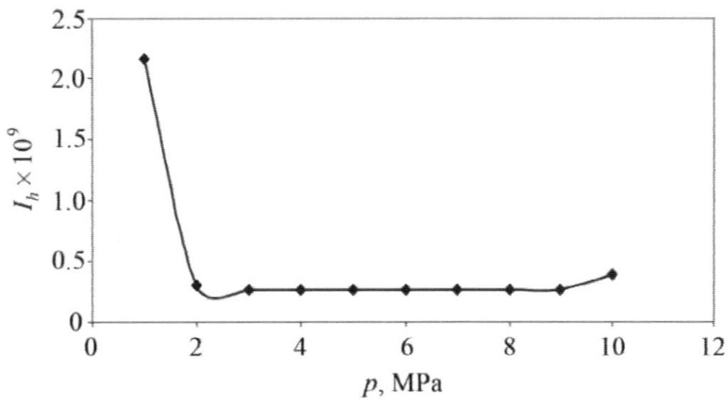

Figure 10. Wear rate I_h of titanium treated by microplasma discharges at a current of ~300 A vs. pressure p applied to the samples.

It could seem paradoxical that the wear resistance of treated samples with the well-developed microrelief turns out to be much higher than that of smooth untreated samples. The matter is that, in standard tribological tests, the friction surfaces of the samples are lubricated with oil. At high pressures applied to untreated samples, oil is forced out of the friction gap, which leads to catastrophic wear and destruction of the sample surface. The microrelief on the sample surface in the form of craters retaining the lubricant prevents extrusion of oil from the friction gap, thereby significantly reducing the friction force and wear rate in a wide range of pressures applied to the samples treated by microplasma discharges.

Biological experiments on the fissiparity of stem cells of bone tissue [10] have shown their good affinity to the microrelief produced by microplasma discharges on titanium surface, which may be useful for medical applications.

Thus, the results of tribological and biological tests demonstrate that the phenomenon of a microplasma discharge excited on a metal surface by a plasma flow opens wide prospects for creating wear resistant materials that can find application in industry, orthopaedics, and stomatology.

Acknowledgments
This work was supported by the Russian Federal Agency for Scientific Organizations (state assignment no. 01200953486).

References

[1] Ivanov V A, Sakharov A S and Konyzhev M E 2008 *Plasma Phys. Rep.* **34** 150
[2] Ivanov V A, Sakharov A S and Konyzhev M E 2016 *Plasma Phys. Rep.* **42** 619
[3] Sakharov A S and Ivanov V A 2016 *Usp. Prikl. Fiz.* **4** 150
[4] Ivanov V A, Jüttner B and Pursch H 1985 *IEEE Trans. Plasma Sci.* **13** 334
[5] Fowler R H and Nordheim L 1928 *Proc. Roy. Soc. London* A **119** 173
[6] Korolev Yu D and Mesyats G A 1982 *Field-Emission and Explosive Processes in Gas Discharges* (Novosibirsk: Nauka) [in Russian]
[7] Oja I, Mere A, Krunks M, Nisumaa R, Solterbeck C-H and Es-Souni M 2006 *Thin Solid Films* **515** 674
[8] Ivanov V A, Konyzhev M E, Kuksenova L I, Lapteva V G and Khrennikova I A 2011 *J. Mach. Manuf. Reliab.* 2015 **44** 384
[9] Ivanov V A, Konyzhev M E, Kamolova T I, Dorofeyuk A A, Kuksenova L I, Lapteva V G, Khrennikova I A and Alekseeva M C 2015 *Usp. Prikl. Fiz.* **1** 47
[10] Dimitrovich D A, Bychkov A I and Ivanov V A 2009 *Prikl. Fiz.* No 2 35

IOP Conf. Series: Journal of Physics: Conf. Series **907** (2017) 012024 doi:10.1088/1742-6596/907/1/012024

Supersonic, subsonic and stationary filaments in the plasma focus

V Ya Nikulin[1,3], S A Startsev[2] and S P Tsybenko[1,4]

[1] Lebedev Physics Institute, 53 Leninsky Pr., Moscow, 119991, Russia
[2] Financial University, 49 Leningradsky Pr., Moscow, 125993, Russia
[3] National Research Nuclear University MEPhI, 31 Kashirskoe Sh., Moscow, 115409, Russia

[4] E-mail: kink@sci.lebedev.ru

Abstract. Filaments in the plasma focus were investigated using a model of plasma with the London current. These structures involve a forward current that flows along the surface of a tangential discontinuity and reverse induction currents in the surrounding plasma, including those that flow over the surface of discontinuity, where the magnetic field reverses its direction. Supersonic filaments demonstrated the capture of plasma by the London current, and in subsonic and stationary filaments, the London current expelled the plasma.

1. Introduction

The plasma focus, which is used in important technological applications to produce rigid radiations and hot plasma flows, is observed to have current filaments with diameters that vary between submillimeter and millimeter ranges [1–3]. The ratio of the diameter of a filament to its length is a small parameter that enables us to consider the filaments as quasicylindrical structures (or as cylindrical structures in the zero approximation with respect to the small parameter).

It has been assumed that filaments can emerge as a result of radiative magnetohydrodynamic instability [4] or their formation can be due to inhomogeneous ionization at the stage of breakdown, and then a multistage ionization enhances plasma density inhomogeneities [5].

In [1], it has been proposed to consider paired plasma vortices to describe the filaments. In [6], stationary filaments were related to helical magnetic field lines. In [7], stationary current filaments were also considered as the states of equilibrium plasma. Then, [8, 9] proposed a model for stationary and moving filaments that involved tangential discontinuity over the surface of which the current flew to induce reverse currents. This paper aims to study the filaments in the plasma model with the London current within the region of parameter values that have not been previously investigated, as well as to discuss plasma capture by the London current and other features of the filaments.

2. Basic equations

The theory is based on dissipationless equations of two-fluid plasma hydrodynamics. These equations have been first used in superconductivity theory [10], which established a relation for the London current $\vec{j} = -e^2 n_e \vec{A} / m_e c$, where $\vec{j}$ is the density of the London current, n_e is the concentration of electrons, and $\vec{A}$ is the vector potential of the electromagnetic field.

We will use the plasma hydrodynamics equations. The characteristic scale of plasma and field inhomogeneity is $L \approx c/\omega_p$, where ω_p is the electron plasma frequency. In this case, the theory of general relativity [11] or variational principles [12] can be used to substantiate the plasma equations. If the hydrodynamic velocity of plasma is much smaller than the current velocity, a transition occurs to the one-fluid model of plasma with the London current.

A simple model of plasma with the London current is a one-fluid model of quasineutral fully ionized plasma [12]:

$$\frac{\partial\rho}{\partial t}+\nabla(\rho\vec{v})=0,\quad \frac{\partial\vec{v}}{\partial t}+(\vec{v}\cdot\nabla)\vec{v}=-\frac{\nabla P}{\rho}-\frac{Ze^2}{2m_e m_i c^2}\nabla\vec{A}^2,\ \nabla\times\nabla\times\vec{A}=-\frac{4\pi Ze^2}{m_e m_i c^2}\rho\vec{A},\quad (1)$$

where Ze and m_i are the charge and mass of the ion, m_e is the mass of the electron, and c is the speed of light. In system (1), pressure $P(\rho)$ is the specified function of density ρ, and $\vec{v}$ is the plasma velocity. Consider the plasma to be adiabatic, i.e., $P=P_0(\rho/\rho_0)^{5/3}$, where P_0 and ρ_0 are the unperturbed pressure and density of the plasma. Besides, from the last equation of the system it follows that $\nabla(\rho\vec{A})=0$. Hence, at $\vec{A}\perp\nabla\rho$, there is the Coulomb calibration for the vector potential $\nabla\vec{A}=0$, which is used below.

In this paper, the filaments are considered as cylindrical structures that can move with a constant velocity D perpendicular to the filament axis. Then, system (1) is reduced to two equations for two independent variables ρ and A_z, where A_z is the component of vector potential in cylindrical coordinates.

Let us use dimensionless variables: $\rho/\rho_0\to\rho$, $\omega_{P0}r/c\to\tau$, $Ze^2A_z^2/2m_e m_i c^2 v_{S0}^2\to a^2$, $D/v_{S0}\to D$. Here, r is the cylindrical coordinate, ω_{P0} and v_{S0} are the electron plasma frequency and speed of sound in unperturbed plasma.

Then, equations (1) take the form of

$$\frac{1}{\tau}\frac{d}{d\tau}\left(\tau\frac{da}{d\tau}\right)=\rho a,\quad a^2=\frac{D^2}{2}\left(1-\frac{1}{\rho^2}\right)+\frac{3}{2}\left(1-\rho^{2/3}\right).\qquad (2)$$

We will seek the solution of system (2) that contains tangential discontinuity. Recall that the combined pressure of plasma and magnetic field is preserved via tangential discontinuity. Note that there is a plasma flow for a cylindrical structure that moves over the tangential discontinuity. Therefore, an approximate relation $P+(\nabla\times\vec{A})^2/8\pi\approx constant$ can be used in this case, provided that the plasma flow over the discontinuity is small.

Let us introduce a coordinate τ_{**} to designate the position of the tangential discontinuity. At $\tau<\tau_{**}$, the solution has only the plasma pressure because the magnetic field vanishes. We will also introduce the coordinate $\tau_*\ (\tau_{**}<\tau_*)$ to designate a discontinuity where the magnetic field reverses its direction. Herewith, all other variables remain continuous such as plasma density $\rho_*=\rho(\tau_*)$. For the dimensionless parameters, the condition of smallness of the plasma flow over a tangential discontinuity will take the form of $D\rho_{**}^{-1}<<1$, where ρ_{**} is the plasma density on the inner side of the discontinuity.

3. Results and discussion

Equations (2) were solved numerically together with the equation for the tangential discontinuity under the condition of smallness of the plasma flow over the tangential discontinuity. The results of the calculations are given in figures 1–4. The solutions are determined by the filament movement velocity D, position of the discontinuity τ_*, where the magnetic field changes its direction stepwise, and density ρ_* at this discontinuity.

First, let us discuss subsonic filaments, which are a particular case of stationary filaments. Here, we present the results for a stationary cylindrical filament (see figure 1 and figure 2). Note that the current flows over the surface of tangential discontinuity in the direction of the axis of the cylindrical system, and induction currents flow around the tangential discontinuity in the reverse direction.

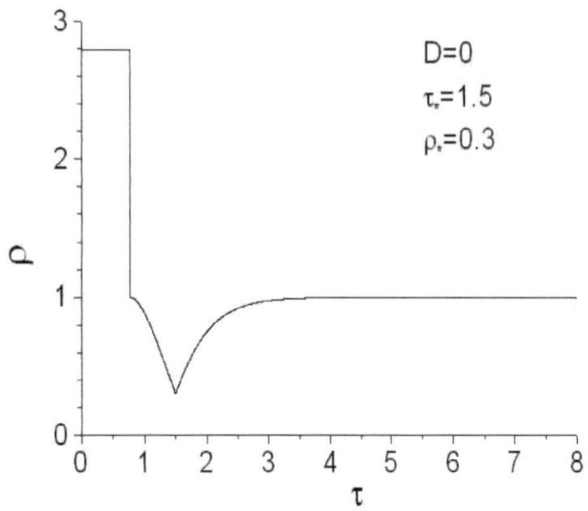

Figure 1. Distribution of plasma density over the radius in a stationary filament.

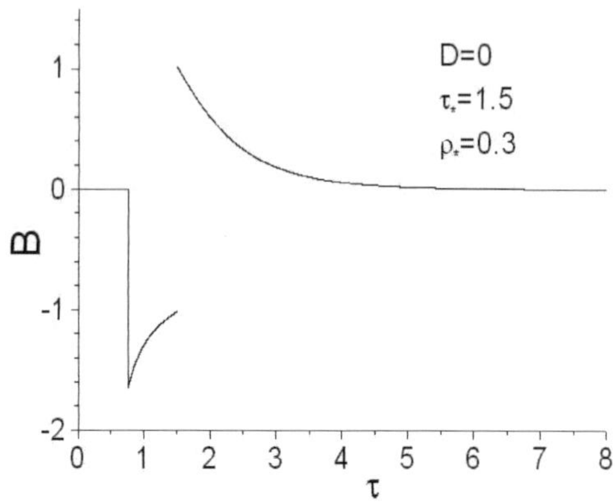

Figure 2. Distribution of magnetic field over the radius in a stationary filament.

Let us assess the parameters of the presented solution. For the concentration of electrons around the cylindrical structure, $n_e \approx 2 \cdot 10^{17}\ cm^{-3}$, the characteristic scale of plasma and field inhomogeneity is $L \approx 15\ \mu m$, i.e., the discussed solutions are attributed to submillimeter structures. The maximal value of magnetic field in the solution in figure 2 is approximately 0.1 MG, if the plasma temperature around the filament is taken to be equal to 100 eV. The current flowing over the surface of the tangential discontinuity is 100 A according to the order of magnitude.

Note also that as ρ_* or τ_* decrease, the maximal values of plasma density and magnetic field increase (the latter increases up to the megagauss magnitude) inside a subsonic or stationary filament [8].

Let us now discuss supersonic filaments. Figures 3 and 4 present the results for a supersonic cylindrical filament, the structure of which also has a tangential discontinuity, and along its surface, the forward current flows to induce reverse currents.

Figure 3 shows a local maximum of plasma density at the coordinate of τ_*. The maximum is related to the maximum of current velocity, which in turn is proportional to the parameter a. The maximum of current velocity is on the coordinate of τ_* for supersonic as well as for subsonic and stationary solutions. From system (2), it follows that the maximum of plasma density in the supersonic solution corresponds to that of current velocity. For the subsonic solution, the minimum of plasma density corresponds to the maximum of current velocity. In other words, the London current captures the plasma in the supersonic solution (figure 3) and expels it in the subsonic or stationary solution (figure 1).

With an increase of ρ_* or a decrease of τ_*, the maximal values of plasma density and magnetic field increase inside the supersonic filament [9].

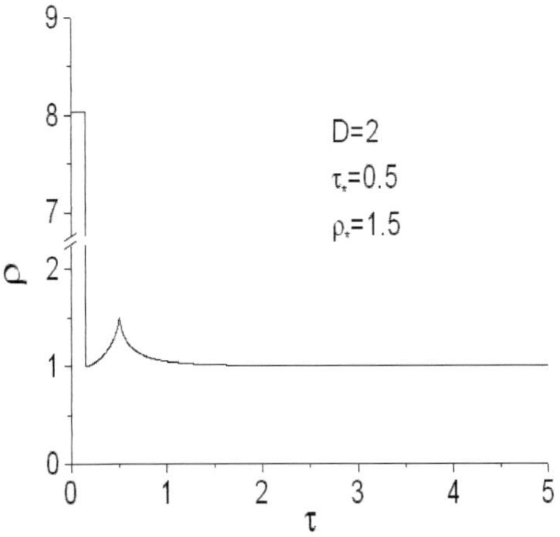

Figure 3. Distribution of plasma density over the radius in a supersonic filament.

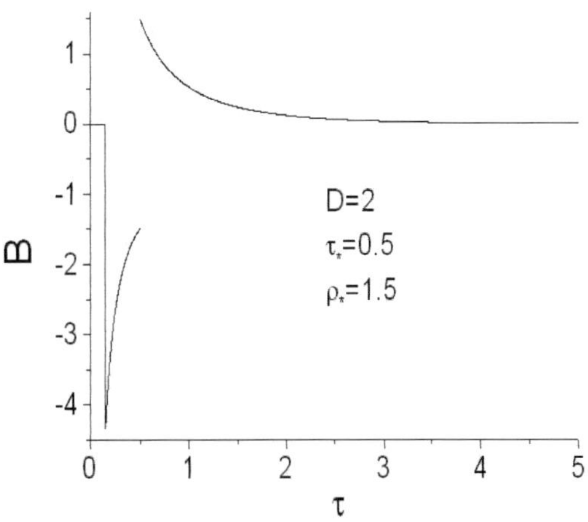

Figure 4. Distribution of magnetic field over the radius in a supersonic filament.

4. Conclusions

A simple model of plasma using the London current provided solutions for the cylindrically moving and stationary filaments. The filaments contain a tangential discontinuity, over which the current flows and induces reverse currents around it. The magnetic fields inside these filaments can reach megagauss values. The obtained solutions for cylindrical filaments proved to be in the submillimeter range.

Acknowledgments

The study was supported by the grant from the Russian Science Foundation (project №16-12-10351).

References

[1] Bostick W H, Prior W, Grunberger L and Emmert G 1966 *Phys. Fluids* **9** 2078
[2] Sadowski W, Herold H, Schmidt H and Shakhatre M 1984 *Phys. Letters A* **105A** 117
[3] Soto L, Pavez C, Castillo F, F Veloso, Moreno J and Auluck S H K 2014 *Phys. Plasmas* **21** 072702
[4] Imshennik V S and Neudachin V V 1987 *Fiz. Plazmy* **13** 1226
[5] Braginskii S I and Vikhrev V V 1976 *Teplofiz. Vysok. Temp.* **14** 254
[6] Nardi V 1970 *Phys. Rev. Lett.* **25** 718
[7] Fadeev F M, Yusupaliev U and Shuteev S A 2004 Proc.15th Intern. Conf. on High-Power Particle Beams "Beams - 2004" (Saint-Petersburg) (Saint-Petersburg: Nauka) p 291
[8] Nikulin V Ya, Startsev S A and Tsybenko S P 2015 *Bull. Lebedev Phys. Inst.* **42** 133
[9] Nikulin V Ya, Startsev S A and Tsybenko S P 2016 *Bull. Lebedev Phys. Inst.* **43** 345
[10] London F 1950 *Superfluids* (New York: Wiley)
[11] Solov'ev L S 1991 *Zh.Eksp.Teor.Fiz.* **100** 455
[12] Nikulin V Ya and Tsybenko S P 1997 *Phys. Scripta* **55** 90

Small-size high-current ion diode with pulsed magnetic insulation of electrons for 500 keV energy

A Isaev[1,2], K Kozlovskiy[1] and E Vovchenko[1]

[1] National Nuclear Research University "MEPhI", 115409, Russia, Moscow, Kashirskoe highway, 31

[2] E-mail: isaev@lenta.ru

Abstract. The work informs about the development of an experimental model of light ion and deuteron accelerator with energy up to 500 keV at a current of approximately 1 kA. The current density of 20 A/cm^2 is reached in pulses with duration of less than 0.5 μs at a repetition rate of 1 Hz. The accelerator applies the magnetic electron insulation technique in the accelerating gap. In addition, an intense laser-plasma ion source is used. The diode features and operation modes are considered using the conical geometry of a spiral line forming the magnetic field.

1. Introduction

Investigations of small-size vacuum laser-plasma diodes, which have been performed in recent years for the purpose of creation of a compact impulse neutron generator [1–3], showed the possibility of effective deuteron acceleration to implement a nuclear reaction D(d,n)^{3}He in a diode with a laser deuterium-containing target on the anode and magnetic insulation of the electron current.

Such diode system has a coaxial geometry of electrons with the inner anode being covered by a hollow cylindrical cathode. The diode accelerates the deuterons arising from an anode target in the direction of the cathode target which is a source on neutrons. In these experiments, the application of a pulsed Nd:YAG laser (neodymium-doped yttrium aluminum garnet), wavelength of λ = 1.06 μm, radiation energy up to W ~ 0.1 J, duration of ~10 ns) and a pulsed high-voltage source with an accelerating voltage of U_{acc} ~ 300 kV, which is made via the Arkadyev–Marx scheme, allowed to produce accelerated deuteron flows with a current of up to 100–150 A [1].

The planning of experiments, which are intended for the augmentation of accelerating voltage up to 500 kV and laser radiation pulse energy up to 1 J, as well as the development and optimization of the diode gap magnetic insulation system, while accounting for the transfer to higher voltages and laser radiation energies, became the natural work progression [3, 4].

The fundamental changes, as compared with previous experiments on the deuteron acceleration in a vacuum diode with laser plasma on the anode, are related to the application of a high-power laser. Another fundamental difference consists in the use of a high-power Arkadyev–Marx generator with the designed shock voltage up to 500 kV. Moreover, the pulsed magnetic insulation is applied in the test experiments instead of the suppression of electronic conductivity by the constant magnetic field.

IOP Conf. Series: Journal of Physics: Conf. Series **907** (2017) 012025 doi:10.1088/1742-6596/907/1/012025

2. Investigation of accelerating ion triode with magnetic insulation

2.1. Experimental determination of the optimal parameters of laser spark sources of deuterons

The deuteron source, which is used in the accelerating diode for neutron generation, is tone of the most important elements of pulsed neutron generator because it mainly defines the threshold characteristics of neutron pulse. In the case of vacuum diode, it can be made based on either vacuum-arc discharge plasma or laser plasma.

Vacuum-arc deuteron source operates most effectively within the microsecond range when the deuteron flight time in diode gap is considerably lower than the accelerating voltage pulse duration. Thereby, a deuteron bunch is formed and accelerated in compliance with the Boguslavskiy–Child–Langmuir model [5]. The model analysis showed that the emission current of deuteron source has a considerable effect on the process of deuteron extraction from plasma. A partial diode locking by a space charge already occurs at the current of ~10 A and a pulse duration of tens of nanoseconds. Besides, the neutron yield and neutron pulse duration are considerably reduced with the growth of emission current from the cathode [6].

Experimental investigations using prototypes showed that the application of spark laser deuteron source, which ensures quick deuteron extraction from plasma, was more effective in the nanosecond range of neutron pulses durations. The use of deuterium plasma, which is created on the anode when the laser is focused on the metal deuteride target, is its distinctive feature.

2.2. Choice of the optimal magnetic field

Different suppression methods of secondary electron emission from cathode, including schemes with magnetic electron insulation in the accelerating diode gap, are used in pulsed ion diodes to increase the energy efficiency of neutron generation.

The application of magnetic insulation in small-size coaxial systems, which are intended for deuteron acceleration and neutron generation, is primarily described in [7]. The electron conduction suppression by the permanent magnet field with the azimuthal symmetry was investigated in [1, 3]. The applications of pulsed magnetic field of spiral line with a current inside the cathode were investigated in [3, 8]. In this case, the critical magnetic field induction in the cathode region of accelerating gap is in the range from 0.2 to 0.6 T for the cathode diameters of 5–10 cm and accelerating voltages of U = 100–500 kV. The experiments show that the electron conduction suppression by the permanent magnet field [1] suffers from grave shortcommings, which occur due to the complex configuration and nonuniformity of the permanent magnet field. This results in the impossibility of axial magnetic field creation with the induction of no less than 0.4 T in the entire space between the cathode and laser plasma front (i.e., plasma anode), which is required for the complete suppression of electron diode conduction at the laser energy of $W > 0.1$ J.

The considered shortcomings manifest themselves to a lesser degree in diodes with the pulsed magnetic insulation. A strong magnetic field is created in the vicinity of the cathode during the current flow via a spiral induction coil, which is located inside. Its generatrix repeats the shape of hollow cathode, which can be made either in the form of a cylinder or a truncated cone. For such geometry, one should take into account that, in the pulsed mode with a high rate of current rise, the inductive current will also affect the process of magnetic field formation by induction coil.

2.3. Schematic of the experiment

The schematic of the experiment is specified in Fig. 1. The insulating magnetic field was generated by the positive-going current pulse. The current generator is made according to the scheme in which a capacitive storage (C_M = 0.25 μF) is discharged through a spiral coil with an inductance of L_M = 0.65 μH. The energy of 25 – 40 J is accumulated in the capacitive storage.

IOP Conf. Series: Journal of Physics: Conf. Series **907** (2017) 012025 doi:10.1088/1742-6596/907/1/012025

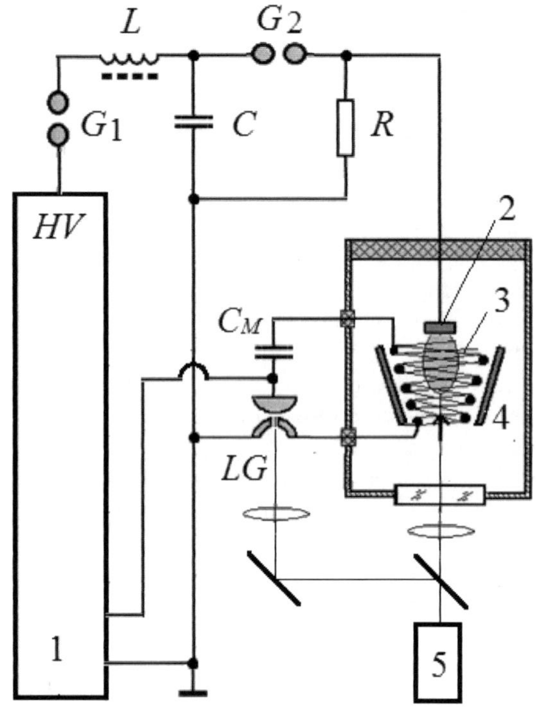

Figure 1. Scheme of the deuteron acceleration experiment in the diode with the laser plasma on the anode and magnetic electron insulation: 1 – Marx generator (L = 400–700 μH, R = 100 kΩ, C = 30 pF, CM = 0.25 μF); 2 – anode; 3 – conic coil; 4 – cathode; 5 – laser.

The characteristic current rise time (i.e., magnetic field) was $\tau_M \approx 400$ ns. A laser switch (LS), which also simultaneously switched the capacity in the first section of the Marx high-voltage pulse generator (HVPG), was applied for switching the storage capacitor C_M. Such scheme ensured the reliable current pulse synchronization (magnetic field generation) with the accelerating voltage pulse. The start of the current pulse coincides with the LS breakdown. Let us note that in these experiments with laser plasma, which are directed at the diode current recording, three processes are synchronized: laser plasma expansion, accelerating voltage formation and generation of the increasing magnetic field. The maximum current that is reached in the inductance coil can be estimated from the energy balance. It is equal to $I_{max} \approx 9$ kA. For this current, the magnetic field at the axis of conical spiral coil is $B_{max} \approx 0.5$–0.6 T. Moreover, already after a quarter of period from the beginning of current rise, the magnetic field is $B \approx 0.7\ B_{max}$. The estimate of the efficiency of magnetic field generation showed that after 300 ns from the start of laser plasma expansion, the current in the conical coil equals to 60% of its peak value.

The circuit consisting of an inductor L = 400–700 μH and a compensating capacitor C ≈ 30 pF (time constant ≈110–150 ns), as well as a spark gap G2, are added to the electric scheme to coordinate the accelerating voltage between the output of the Marx generator and the diode load with the processes of acceleration in the diode gap and the increasing magnetic field. In this case, the total delay of voltage pulse relative to laser pulse equals to $\tau_{HV} \approx 200$ ns. Inductor is a 20-turn cylindrical spiral, where a ferrite core, M400HH, (70-mm diameter, 500-mm length) is inserted. The spark gap G_1 separates the LC circuit from the charging voltage of the Marx generator.

A Nd:YAG laser, which generates a radiation pulses with the wavelength of 1.06 μm, energy ≤0.85 J and duration ≈10 ns in the high-Q resonator mode, was used to obtain laser plasma. The current in the experiment was measured by the Rogowsky coils.

Under our experimental conditions, the upper limit of accelerating voltage was determined only by the maximum voltage that is generated by the Marx generator. In the absence of laser radiation on the anode, the vacuum gap breakdown did not progress. If the laser plasma was present on the anode, the initial stage of the discharge rapidly turns into the high-current spark stage, which was stipulated by the development of electronic current and conductivity rise in the gap between the plasma anode and cathode. In turn, this led to the reduction of accelerating voltage. However, the application of pulsed insulating magnetic field allowed a considerable suppression of electron current. Moreover, the current in the spiral coil, which was equal to ~5 kA, prevented the breakdown between the cathode and anode during the first 0.5 μs from the initiation of laser pulse. The current of up to 1 kA at the ultimate accelerating voltage of up to 400 kV was recorded in the triode with magnetic insulation at the radiation energy on the laser target of 0.75 J.

The abovementioned information is required for the theoretical estimate of possible neutron yield in the considered diode system. The neutron yield per pulse can be determined using the following equation:

$$Q(W) \approx \frac{n}{e}\int_0^{\tau} dt I_D(W,t) \int_0^{eU(t)} dE \frac{\sigma(E)}{F(E)},$$

where n is the concentration of reagent nuclei in the neutron forming target, e is the elementary electric charge, τ is the pulse duration of the deuteron current, $U(t)$ is the time dependence of the accelerating voltage, $F(E)$ is the deuteron energy loss per unit of target length, and $\sigma(E)$ is the micro-section of the nuclear reaction.

3. Conclusions

The current of approximately 1 kA at the ultimate accelerating voltage up to 500 kV was achieved at the laser radiation energy on the target of 0.75 J in the diode with magnetic insulation. The current in spiral coil, which was equal to 5 kA, prevented the breakdown between the cathode and anode during the first 0.5 μs.

Calculations of the optimal apex angle α of the conical cathode and spiral coil near the cathode gave the value tgα ≤ 0.5. Moreover, at this angle, the deuteron yield from the anode plasma is maximal.

The application of pulsed magnetic insulation using the magnetic field generated by a spiral coil with the current flowing through it showed the complete absence of breakdowns between triode electrodes. This scheme has advantages as compared to the permanent magnet system, although it is less compact than the latter.

The optimal magnetic field induction in the gap between the plasma anode and cathode with a pulsed solenoid coil was found to be in the range from 0.5 to 0.8 T.

At the optimal magnetic field, the designed accelerating ion triode makes it possible to create deuteron bunches with energy of approximately 500 keV with currents of approximately 1 kA in the systems with the size of less than 10 cm.

Calculations that are based on experimental data on the obtained bunches of accelerated light ions, show that the neutron yield of nuclear reaction of $T(d,n)^4He$ can reach 10^{11} neutr./pulse.

References

[1] Shykanov A E, Vovchenko E D, Kozlovskiy K I and Shatokhin V L Diode for acceleration of hydrogen nuclides with electronic conduction suppression by inner ring magnet 2015 *Letters to Russian Journal of Applied Physics* **41** 10 104-110

[2] Shykanov A E, Vovchenko E D and Kozlovskiy K I Neutron generation in plasma diode with electron isoltation by permanent magnet field 2015 *Nuclear energy* **119** 4 210-215

[3] Didenko A N, Shykanov A E, Kozlovskiy K I, Shatokhin V L and Ponomarev D D Small-size plasma diodes with magnetic insulation for neutron generation 2014 *Plasma Physics Reports* **40** 11 1025-1034.

[4] Kozlovskiy K I, Ponomarev D D, Ryzhkov V I, Tsybin A S and Shykanov A E Model investigational study of small-size neutorn generator with impulse magnetic insulation 2012 *Nuclear energy* **112** 3 182-184

[5] Bogdanovich B Y, Nesterovich A V, Shykanov A E et al Remote radiation monitoring with linear acclerators of charged particles. Vol. 1. Linear accelerators for generation of braking radiation and neutrons 2009 M.: Energoatomizdat p 272

[6] Didenko A N, Shykanov A E, Raschikov V I, Ryzhkov V I and Shatokhin V L Generation of nanosecond neutron impulses in vacuum accelerating tubes. Technical Physics 2014 *Russian Journal of Applied Physics* **84** 6 119-125.
[7] Bespalov D F, Kozlovskiy K I, Tsybin A S and Shykanov A E Pulse neutron tube: A.c. 1979 766048 USSR
[8] Gulko V M, Dydychkin V N, Kozlovskiy K I, Kolomiets N F, Mikhaylenko B V, Shykanov A E and Yakovlev K I Pulse neutron tube: A.c. 1985 1468275 USSR

IOP Conf. Series: Journal of Physics: Conf. Series **907** (2017) 012026 doi:10.1088/1742-6596/907/1/012026

Laboratory simulations of astrophysical jets: results from experiments at the PF-3, PF-1000U, and KPF-4 facilities

V I Krauz[1,7], V V Myalton[1], V P Vinogradov[1], E P Velikhov[1], S S Ananyev[1], S A Dan'ko[1], Yu G Kalinin[1], A M Kharrasov[1], Yu V Vinogradova[1], K N Mitrofanov[2], M Paduch[3], R Miklaszewski[3], E Zielinska[3], E Skladnik-Sadowska[4], M J Sadowski[4], R Kwiatkowski[4], K Tomaszewski[5] and D A Vojtenko[6]

[1] NRC Kurchatov Institute, Moscow, Russia
[2] TRINITI, Troitsk, Moscow, Russia
[3] IFPiLM, Warsaw, Poland
[4] NCBJ, Otwock-Swierk, Poland
[5] ACS Ltd, Warsaw, Poland
[6] SUE "SFTI", Sukhum, Abkhazia

[7] E-mail: krauz_vi@nrcki.ru

Abstract. Results are presented from laboratory simulations of plasma jets emitted by young stellar objects carried out at the plasma focus facilities. The experiments were performed at three facilities: the PF-3, PF-1000U and KPF-4. The operation modes were realized enabling the formation of narrow plasma jets which can propagate over long distances. The main parameters of plasma jets and background plasma were determined. In order to control the ratio of a jet density to that of background plasma, some special operation modes with pulsed injection of the working gas were used.

1. Introduction

Laboratory simulations of astrophysical phenomena are one of the priority directions in the scientific research [1]. Substantial progress in modelling the astrophysical jets has been achieved using high-power lasers [2] and Z-pinch facilities [3]. Recent studies have shown that the plasma focus (PF) facilities can be successfully used for modelling plasma jets emitted by young stellar objects (YSO). It was shown that plasma jets generated during the PF discharges can propagate over distances which are considerably larger than the transversal dimensions of these jets [4]. A plasma jet with a density of $\geq 2 \times 10^{17}$ cm^{-3} and a temperature of a few electron-volts can propagate through background plasma with a density of $\geq 2 \times 10^{16}$ cm^{-3} [5-6]. It was also shown that, in this case, an axial current appears which generates the toroidal magnetic field confining the plasma jet [7-8]. Problem of stability is one of the important issues in studies of plasma jet dynamics.

Currently, an international scientific collaboration is being developed which involves studies at three large PF facilities, notably, the PF-3 facility at the NRC "Kurchatov Institute" in Moscow (Russia), the PF-1000U facility at the IFPiLM in Warsaw (Poland), and the KPF-4 facility at the SUE "SFTI" in Sukhum (Abkhazia). The paper presents a short review of the main results, which were recently obtained at these facilities.

IOP Conf. Series: Journal of Physics: Conf. Series **907** (2017) 012026 doi:10.1088/1742-6596/907/1/012026

2. The PF-3 facility

At the PF-3 facility, at the stationary filling of the experimental chamber with the working gas , the modes of operation with the formation of narrow plasma jets were obtained (see Fig. 1). Generally, such modes are achieved using heavy and strongly-radiating gases. In the course of propagation over a distance of about 100 cm, the transversal dimensions of a plasma jet head do not exceed several centimetres (see Fig. 1a). The observed radial distribution of the toroidal magnetic field is consistent with the distribution generated by an axial current of about 3-10 kA, which flows in the paraxial region with a radius of 1-2 cm (see Fig. 1b).

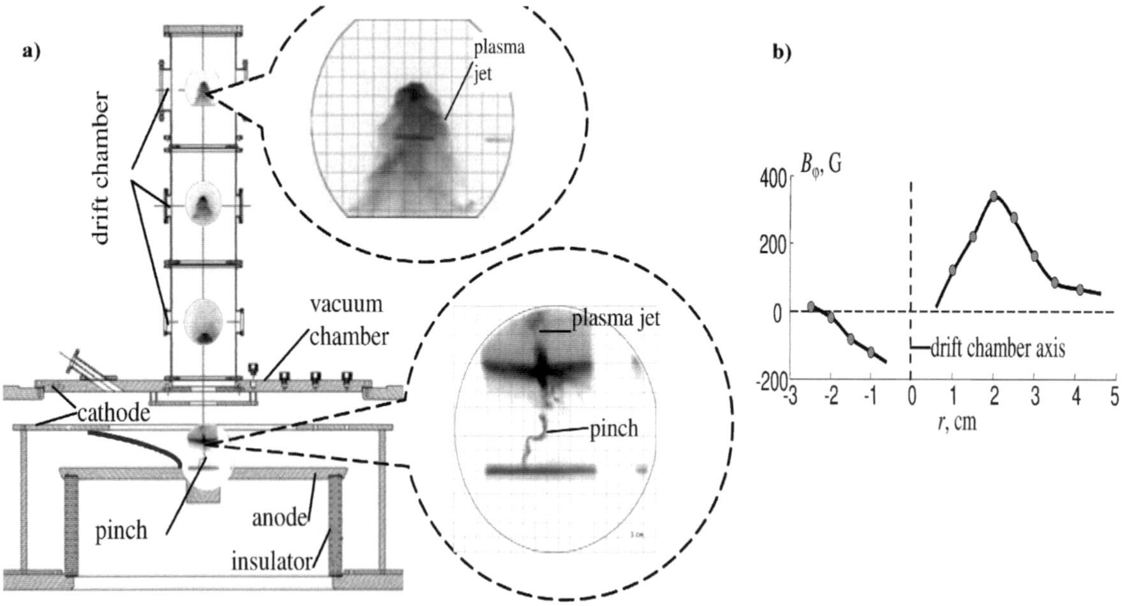

Figure 1. a) Schematic of the PF-3 facility and images of the plasma jet taken in the stage of jet formation (bottom) and at a distance of 95 cm from the anode surface (top); the exposure was 12 ns. b) Radial distribution of the toroidal magnetic field at a distance of 35 cm from the anode end.

At a periphery of the plasma stream one can observe magnetic field structures induced by a flow of return currents.In spite of a strong damping of the magnetic field in the course of the jet propagation, for the PF discharges in neon, the plasma jet head remains compact which can be due to the radiative cooling of plasma.

3. The PF-1000U facility

The experiments at the PF-1000U facility, as well as at the KPF-4 facility, were mainly aimed to create different initial distributions of the working gas density. This makes it possible to manage changes in the conditions of the plasma jet propagation in the background plasma.

The PF-1000U facility, which is the Mather-type facility, can operate in the modes with additional puffing of the working gas into the paraxial region of the discharge. The experimental setup is shown in Fig. 2.

The experiments were carried out at discharge energy of 170 kJ, and at the stationary filling of deuterium with an initial pressure of 1.2 hPa. To produce higher gas density in the pinch region, different gases were used for the additional gas puffing: deuterium, helium, neon and their mixtures. Gases were injected through a pulsed valve which was opened about 2 ms before the discharge initiation.

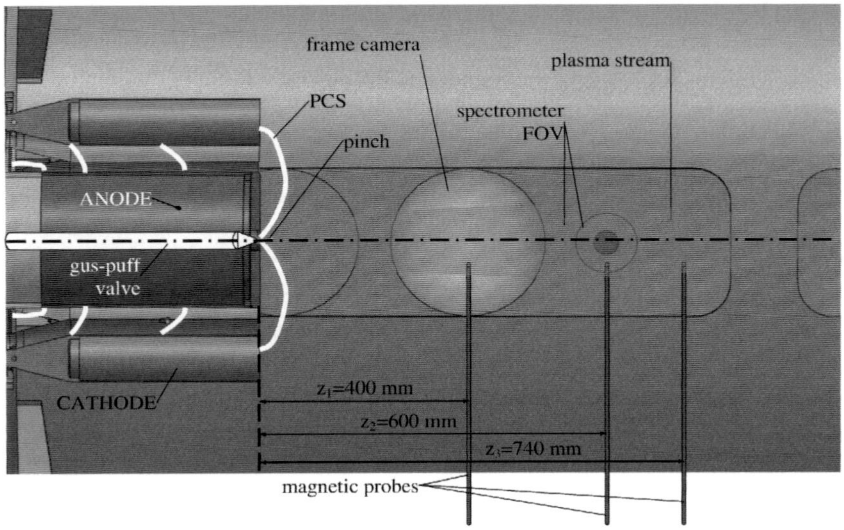

Figure 2. Setup of the PF-1000U experiment performed with the additional gas puffing.

The plasma jet structure was studied at a distance of 40 cm from the anode end using an optical single-frame camera with an exposure of 10 ns. It was observed that the plasma jet structure considerably differed for discharges with the stationary and combined (stationary + additional) gas puffing (see Fig. 3). In the second case, the compact plasma structures with dimensions of several centimetres formed. Signals of the magnetic probes showed that the axial current flowed through these plasma structures. A complicated funnel-like shape of the shock front might be caused by the reverse currents at a plasma jet periphery.

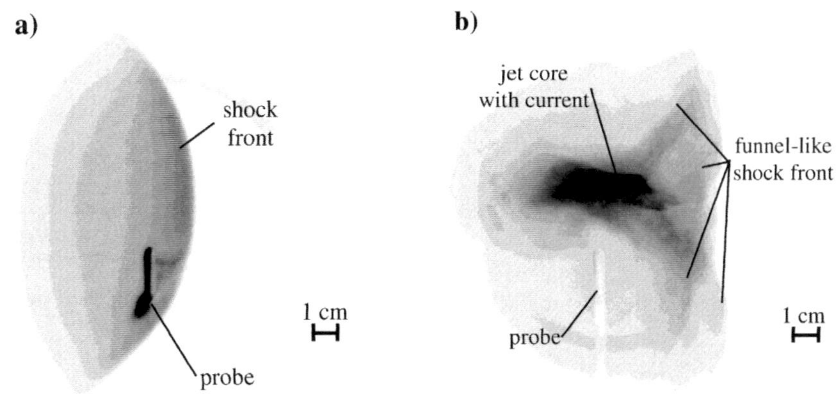

Figure 3. Frame images of a plasma jet taken at a distance of 40 cm from the anode end: a) the stationary filling of deuterium at the initial pressure of 1.2 hPa; b) the combined filling of deuterium at the initial pressure of 1.2 hPa with the additional injection of a gas mixture (deuterium (75%) and neon (25%)). The magnetic probe which recorded the azimuthal magnetic field is also seen in the images.

The magnetic field damping due to the dissipation of the axial current was observed in the course of propagation of the plasma jet along the axis. In the case of the stationary deuterium filling, only the shock wave front is seen in the image. However, the magnetic probe recorded the appearance of the

IOP Conf. Series: Journal of Physics: Conf. Series **907** (2017) 012026 doi:10.1088/1742-6596/907/1/012026

magnetic field behind the front of the shock wave, which is induced by the axial current. Evidently, the plasma parameters achieved in this case were insufficient for the visualization of the jet core.

The plasma density at a distance of 57 cm from the anode end was estimated from the Stark broadening of spectral lines. It occurred to be (0.4–3.7) x 10^{17} cm^{-3} and depended on the initial gas distribution and delay time between the spectrum registration and the plasma jet generation. The electron temperature of the jet plasma amounted to about 5 eV, and the density of background plasma was about 1.5 x 10^{15} cm^{-3} [6].

4. The KPF-4 facility

The KPF-4 facility can operate in the mode with the pulsed injection of the working gas, which differs from that used at the PF-1000U facility, because it provides an increased gas density near the insulator surface. The working gas is puffed using a fast valve equipped with 36 nozzles, which are distributed symmetrically around the cathode disk and directed towards the insulator surface. The gas is puffed after preliminary pumping the experimental chamber down to the pressure of about 10^{-2} hPa. Measurements performed using electrodynamic pressure sensors showed that the optimal time delay between triggering the gas valve and the discharge initiation is 5-6 ms. At this delay time, the pressure near the insulator and in the acceleration channel amounts to ≥ 1 hPa, which is sufficient for the development of the discharge, although the gas pressure in the region of the plasma jet propagation is low (much lower than 1 hPa). Thus, at some distance from the anode end, the conditions of high "density contrast" are achieved (i.e. the ratio of the plasma jet density to the background gas density increases). In these experiments as opposed to the experiments with stationary filling of the vacuum chamber, the plasma jet propagated without considerable deceleration (see Fig. 4).

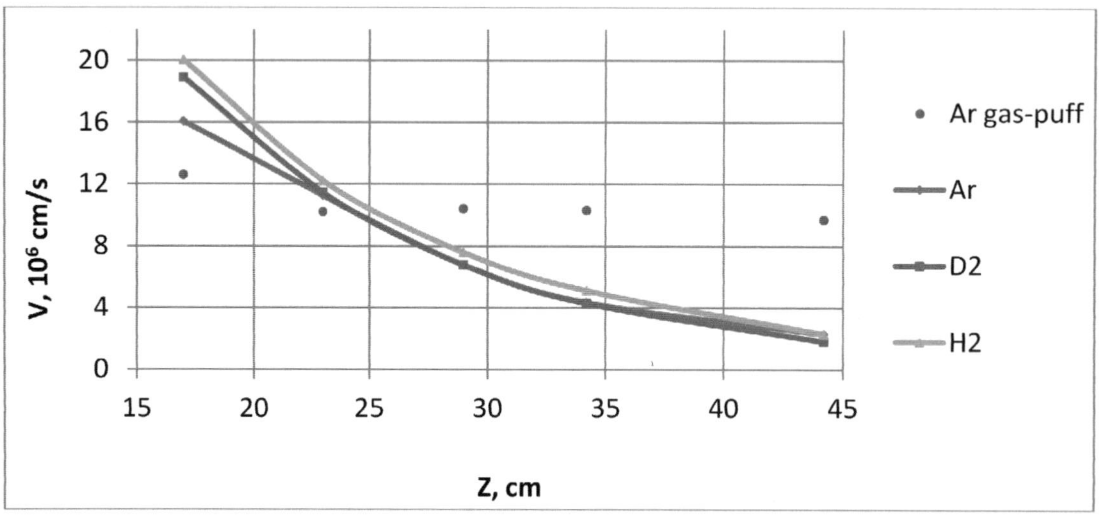

Figure 4. Deceleration of the plasma jet in the KPF-4 facility operating under various conditions: at the stationary pressure and with additional gas puffing.

In experiments on localization of the magnetic field inside the plasma jet, a comparison between the magnetic probe signals and optical streak images was performed. It was shown that the magnetic field captured by the plasma jet is localized in the regions with weak optical emission from plasma, i.e., in the so-called "magnetic bubbles".

5. Conclusions

In the experiments performed at three PF facilities with different configurations and systems of the working gas supply, we observed the similar processes of plasma jet generation in a wide range of experimental conditions. In laboratory experiments, plasma jets with velocities up to ~10^7 cm/s and the Mach numbers of M = 2–10 were observed. These parameters are comparable to those of the astrophysical jets. The obtained data made it possible to compare the main dimensionless parameters. The hydrodynamic and magnetic Reynolds numbers achieved in the experiments, as well as the Pécklet numbers, are much higher than 1, similarly to these parameters of the astrophysical jets. It may be the reason for using the PF facilities for laboratory simulations of astrophysical jets, because the MHD equations (which describe laboratory and astrophysical plasma jets) can be scaled up in space and time.

An important feature of the described experimental systems is the possibility of studying stability of the plasma jets in the course of their propagation over large distances at different parameters of background plasma.

Acknowledgments
This work was supported in part by the Russian Foundation for Basic Research (project nos. 14-29-06085 and 17-52-40008), and the International Atomic Energy Agency (project no. RC-19253).

References
[1] Remington B A, Drake R P and Ryutov D D 2006 *Rev. Modern Phys.* **78** No 3 755-807
[2] Suzuki-Vidal F et al 2012 *Physics of Plasmas* **19** 022708
[3] Albertazzi B et al 2014 *Science* **346** 325-328
[4] Krauz V, Myalton V, Vinogradov V et al 2015 *42nd EPS CPP* Lisbon Portugal **39E** 4.401
[5] Ananyev S S, Dan'ko S A, Myalton V V et al 2016 *Plasma Phys. Reports* **42** 269–277
[6] Skladnik-Sadowska E, Dan'ko S A, Kwiatkowski R et al 2016 *Phys. Plasmas* **23** 122902
[7] Mitrofanov K N, Krauz V I, Myalton V V et al 2014 *JETP* **119** 910-923
[8] Mitrofanov K N, Krauz V I, Myalton et al 2017 *Astronomy Reports* **61** No. 2 138–152

IOP Conf. Series: Journal of Physics: Conf. Series **907** (2017) 012027 doi:10.1088/1742-6596/907/1/012027

Development of precursors recognition methods in vector signals

V G Kapralov[1,3], V V Elagin[1], E G Kaveeva[1], L A Stankevich[1], M M Dremin[2], S V Krylov[2], A E Borovov[1], H A Harfush[1] and K S Sedov[1]

[1] SPbPU, Saint-Petersburg, RF
[2] NRC «Kurchatov Institute», Moscow, RF

[3] E-mail: kapralov@phtf.stu.neva.ru

Abstract. Precursor recognition methods in vector signals of plasma diagnostics are presented. Their requirements and possible options for their development are considered. In particular, the variants of using symbolic regression for building a plasma disruption prediction system are discussed. The initial data preparation using correlation analysis and symbolic regression is discussed. Special attention is paid to the possibility of using algorithms in real time.

1.Introduction

As the design of plasma machines becomes more complex and the understanding of processes that occur in them becomes more profound, and due to the increase in productivity of computing systems and the performance of measuring and actuating devices, as well as development of new mathematical methods and algorithms, the task of predicting the behavior of plasma and predicting any threshold events in plasma discharge arises more often, including the conditions of processing multichannel data in real time.

Thus, a typical problem for tokamaks is the prediction of plasma disruption. Plasma disruption is a transient event during which the plasma discharge ceases. During the plasma quench, all of the energy that is stored in it during the discharge time leaves the plasma cord, which leads to large loads on the structural elements of the tokamak and the surface of the first wall of the vacuum chamber. In addition, during the plasma quench, runaway electron beams can be formed, which have entered into the continuous acceleration conditions. These beams, upon reaching the chamber wall, can locally melt its surface. To reduce the negative effects of plasma disruption, a number of methods and systems have been developed to mitigate the quench such as impulse massive gas injection. However, for the operation of such systems, up to 10 ms is needed. Thus, it is necessary to predict the time of disruption and generate a trigger signal with the required anticipation. Therefore, a disruption prediction system is used that calculates the probability of disruption after a certain period of time in the future. If a certain threshold probability value is exceeded, a signal must be generated to trigger a controlled quench mitigation system to prevent the formation and provide suppression of runaway electron beams. The longer is the time interval for which a prediction of disruption is made, the simpler is the technical implementation of the systems that are involved in the control of the quench, and the wider range of possible systems that can be used to neutralize the undesirable consequences of the predicted disruption.

Based on the previous results on this subject [1], for ITER (using database IGDBTH4V6B), symbolic regression methods and genetic algorithms are used to form the scaling of temperature threshold for the transition into improved confinement mode. In this mode, in a certain layer of plasma column, the transport coefficients of particles and heat decrease, which leads to the formation of a transport barrier and increases the plasma density and temperature inside it. In [2], the APODIS (Advanced Predictor Of DISruptions) system is considered, which was used on the JET tokamak for predicting plasma disruptions. A sufficient reliability of prediction with a small number of false and missed alarms is demonstrated, even for the highly noisy data. In addition, the methods for forming training data sets to detect inaccurate experimental data in real time are proposed. For APODIS, two synthetic signal generators have also been developed [3], which allow interpolation of data gaps. This allows the disruption prediction system to be operational if there are inaccurate or missed data.

The purpose of this paper was to consider approaches for taking into account during the formation of precursors not only explicit dependencies in the original data but also hidden dependencies by simply including in the number of considered parameters the time samples, which are registered separately for each of the processes with a different frequency.

2. Data selection

The formation (selection) of input data of the parameters both of the plasma during the discharge and the T-10 tokamak itself is an important stage during the construction of a system for processing vector signals, especially in real time. At this step, it is necessary to exclude interdependent signals to reduce the demand on hardware performance by discarding duplicate information at the early stage of system construction.

Primary processing was performed using a correlation analysis. As a result, a set of input signals was formed, which was an order of magnitude smaller than the set of all detected signals in the tokamak database. The primary set of parameters consisted of the total plasma current, electron temperature and density at the center of the plasma column, and the effective charge of the plasma. The correlation coefficients between these parameters are presented in Table 1.

Table 1. Correlation coefficients between selected plasma parameters

Parameters	Plasma current I_p	Electron temperature T_e	Electron density n_e	Effective charge Z_{eff}
I_p	1.00	0.16	-0.32	-0.33
T_e	0.16	1.00	0.41	-0.21
n_e	-0.32	0.41	1.00	0.21
Z_{eff}	-0.33	-0.21	0.21	1.00

3. Drift of tokamak parameters

During the development of a system for predicting plasma disruptions, it is possible to consider various options of its functioning. It is desirable to consider these options during the process of system development and training.

The first such case takes into account the slow changes of machine parameters with time. The system is trained using data that was received in the past. The data are applied to signals that have not yet been registered at the time of training. Thus, for the time since learning, the coefficients of functional dependencies could have changed somewhat, for example, due to the change in surface quality of the first wall of the vacuum chamber. Moreover, such drift of parameters can be observed within the same day of tokamak operation. The first morning shots are performed at one vacuum conditions and recycling, and later, the situation may change.

The second case that should be considered when constructing a prediction system is the possibility of taking into account the significant events that occur in the tokamak, both during the discharges and in the intervals between them. Such events include substandard shots, sudden change in vacuum conditions, and switching on of active systems such as additional heating or injection. In this case, it is necessary to switch the disruption prediction system to pre-configured settings for the new operating mode. Preferably, this should be performed automatically.

The third case that needs to be taken into account during the development and training of the disruption prediction system is the change in the set of available signals. In this situation, the system can be switched to either work on available signals or include the calculation of missing data on available experimental data in real time.

To take into account the described situations, we introduce a number of parameters - firstly, the temporal and phase characteristics of the plasma machine. The absolute time t_U, which has been proposed to count in milliseconds since 01/01/1970 (the beginning of the UNIX era), is convenient for programming, data acquisition and automation purposes. The accounting of cyclic changes in tokamak parameters is suggested to be considered using the time from the start of the cycle t_i^c, duration of the cycle T_i^c, duration of the tokamak idle time before the start of the cycle T_i^s, and phase of the cycle $\theta_i^c = t_i^c / T_i^c$. Index i specifies the type of cycle: 1 - experimental campaign, 2 - working week, 3 - working day, 4 - preparation and execution of one shot, 5 - plasma discharge itself.

For these time and phase characteristics, as well as the previously selected parameters given in Table 1, a correlation analysis was performed. Most of the coefficients in the correlation matrix were close to zero values.

Symbolic regression was used to extract the functional dependencies of the selected parameters on time and phase characteristics [4]. In addition to the basic arithmetic operations, the power x^a and trigonometric *sin(cx + d)* functions were considered as permissible functional dependencies of the variable x.

In this case, the dependences were found, the most interesting of which are the dependence of electron temperature on the phase θ_3^c of the cycle that corresponds to the working day

$$T_e = T_{e0}\left(1 + 0.04\sin\left(4\pi\theta_3^c - \pi/2\right)\right) \quad (1)$$

and the effective charge of plasma from the phase θ2c of the cycle that corresponds to the working week

$$Z_{eff} = Z_{eff\,0}\left(1 - 0.06\theta_2^{c\,2}\left(1 - \theta_2^c\right)\right) \quad (2)$$

Here, the normalization of values of the electron temperature and effective charge is carried out according to the data of the third conditioned discharge in the corresponding cycle.

Thus, in the disruption prediction system using temporal parameters, it is available to consider the influence of other parameters that are not explicitly taken into account because of the difficulties in measuring or calculating them.

4. Modeling

To test the suggested technique for taking into account the drift of tokamak parameters, a set of shots was taken from the tokamak T-10 database, which was processed with and without the time parameters θic that were introduced in the previous paragraph. The results of the simulation are presented in Table 2.

Four models were considered, which differed in the sets of parameters that were taken into account. A basic set of parameters, including plasma current, electron temperature and density at the center of plasma discharge, were taken into account in all four models. The effective charge, the calculation of which complicates the application of model in real time, was considered only in models 2 and 4. Finally, the time parameters of phases of the working day and week were taken into account in models 3 and 4, according to expressions (1) and (2), respectively. The data in Table 2 are averaged over a set of 8 shots, which were used as input data for every model.

Table 2. Values of the probability of disruption for different sets of parameters and intervals of disruption prediction

Model index	Inclusion of the basic set of parameters	Inclusion of effective charge in the model	Inclusion of time parameters in the model	Time before disruption when probability exceed 70%	Probability of disruption at 10 ms prior to it	Probability of disruption at 5 ms prior to it
	Ip, Te, ne	Zeff	θ_2^c, θ_3^c	t_{70}, *ms*	p_{-10}	p_{-5}
1	+	-	-	4.2	0.44	0.56
2	+	+	-	4.8	0.55	0.62
3	+	-	+	4.8	0.53	0.64
4	+	+	+	5.1	0.56	0.71

It can be seen in Table 2 that the addition of any of the parameters under consideration improves the disruption prediction result. However, even when all parameters are considered, the probability of 70%, which is sufficient to start the quench mitigation system, reached 5 ms before the breakdown only for model 4.

5. Conclusions

During the process of developing and training a system for disruption prediction, one must take into account the slow drift of tokamak parameters with time both for a long time and during a working day. It is also important to take into account the changes in situation after disruptions or any rare but significant events such as change in the materials of a part of the first wall. It should be possible to automatically switch the modes of operation of the prediction system when active systems, such as additional heating or pellet injection, are switched on.

The disruption prediction system should allow to adaptively change the prediction time interval, depending on the quality of input data and devices that are switched on by the trigger. In addition, there should be the possibility of removing invalid data and interpolating data gaps.

The use of correlation analysis for the selection of parameters and the method of symbolic regression for the formation of functional dependencies during the construction of a system for plasma disruption prediction have been shown. Similar data processing schemes are also used in other areas, for example, when processing EEG signals [5].

Acknowledgements

The authors are thankful to the staff of the T-10 tokamak for the provided data and support. The work was carried out with the partial support from the State Corporation ROSATOM, contract No.H.4x.241.9B.17.1011.

References

[1] Peluso E, Murari A, Gelfusa M and Gaudio P 2014 *Plasma Physics and Controlled Fusion* **56** 114001

[2] Moreno R et al 2014 *Plasma Physics and Controlled Fusion* **56** 114003

[3] Rattá G A, Vega J and Murari A 2014 *Plasma Physics and Controlled Fusion* **56** 114004

[4] Awange J L and Pal B 2016 Symbolic Regression *Geospatial Algebraic Computations* (Berlin Heidelberg: Springer-Verlag) part 1 chapter 11 pp 203-216 DOI:10.1007/978-3-319-25465-4_11

[5] Stankevich L A et al 2016 EEG pattern decoding of rhythmic individual finger imaginary movements of one hand *Human Physiology* **42** 32

IOP Conf. Series: Journal of Physics: Conf. Series **907** (2017) 012028 doi:10.1088/1742-6596/907/1/012028

Coulomb explosions in plasma

Zh B Seksembayev[1], O S Bayakhmetov and S K Sakhiyev

L N Gumilyov Eurasian National University, Astana, Kazakhstan

[1] E-mail: jandos_s90@mail.ru

Abstract. Dense and uniformly distributed in volume ions due to uncompensated electrical charge can "explode", thus creating a directed beam of high-energy particles. The obtained energy during the Coulomb "explosion" can reach high values. It was determined that the energy values and time of the explosion depend on a number of particle system parameters such as charge, density, and dimensions.
In this paper, the calculations of the "explosion" process using a dynamical approach for different configurations are presented: particle–particle, particles in cylindrical and spherical volumes. The "linearization" method is used for calculating the explosion time, particle velocity and energy. The correctness of the applied approach is shown in comparison with the traditional one. The calculations for the configurations are made using the necessary approximations. Numerical calculations are carried out for the case of the double size change.

1. Introduction

Dense and uniformly distributed in volume ions due to uncompensated electrical charge can "explode", thus creating a directed beam of high-energy particles. In a series of papers [1–3], it has been shown that the effect of target's Coulomb "explosion", which is used as an accelerator prototype, plays an important role in particle acceleration. Sometimes the energy of the accelerated using this approach particles reaches values in the MeV range. For a more effective use of such event in the setups, one should provide the corresponding conditions. It is possible to obtain energies that are sufficient to trigger a nuclear fusion process, which is a good aspect for the application of the effects in setups and techniques.

The definition of the Coulomb "explosion" process parameters (dimension, time, velocity, and energy of the particles) is an important problem because the knowledge of the exact parameters is required for constructing setups with fusion process implementation and for other applied problems. Explosion time management is very important for the effective use of the phenomenon effects. Thus, by knowing the characteristic times of "explosions", it is possible to select the optimal time synchronization of the associated processes, which would increase the efficiency of devices used (for example, in [2], the characteristic times allow us to estimate the effect of laser on a bound cluster that consists of a multitude of atoms). The knowledge of velocities at the output during the "explosion" will allow to calculate energy and use beams for various technical problems. In addition, the knowledge of dimensions will allow to design an optimal configuration scheme of the setup. Proceeding from these justifications, the purpose of this paper is an analytical calculation of the above mentioned "explosion" processes.

The calculations, which are presented in this paper, were carried out for the systems of similar ions with identical charges. Nevertheless, the results are also applicable to systems of dissimilar particles

IOP Conf. Series: Journal of Physics: Conf. Series **907** (2017) 012028 doi:10.1088/1742-6596/907/1/012028

with an uncompensated total charge value, partially ionized particles, and can be generalized to different charges.

There are many terms of creation and applications of Coulomb "explosion" phenomenon. The parameters of interaction of two free point charges are calculated in Part 2. The calculation of parameters of some fundamental cases is presented in Parts 3 and 4. The times for double size change in selected direction, average velocity and achievable velocity for this interval are calculated for each configuration. These values are completely applicable as characteristic [2].

2. System with two free point-like identical charged particles

It is known that there is a Coulomb force between charged particles, and it is defined as $F_c = q_2 E = k\dfrac{q_1 q_2}{r^2}$, where q_1 is the charge that creates the electrical field E at the point where charge q_2 is situated, and r is the distance between the charges.

Let us consider a system of two identical free point-like charges $q_1 = q_2 = q$ with a mass m_0, which are initially at rest at a distance L_0 between them (Fig.1). We "linearize" the distance between particles at any time t, i.e., we represent a linear dependence of the distance as a function of time, in which the constant coefficient is equal to the average velocity that corresponds directly to this interval. ($L \sim \upsilon_{av} t$). The equation of motion for each charge that moves separately with respect to the symmetry axis (dashed line) is $F_c = k\dfrac{q^2}{[2(L_0/2+\upsilon_{av}t)]^2} = m_0 \dfrac{d\upsilon}{dt}$. By solving and integrating with respect to the time between $t_0 = 0$ and t, we obtain at an arbitrary moment $\upsilon = \upsilon_0 + k\dfrac{q^2 t}{m_0 L_0 (L_0 + 2\upsilon_{av} t)}$.

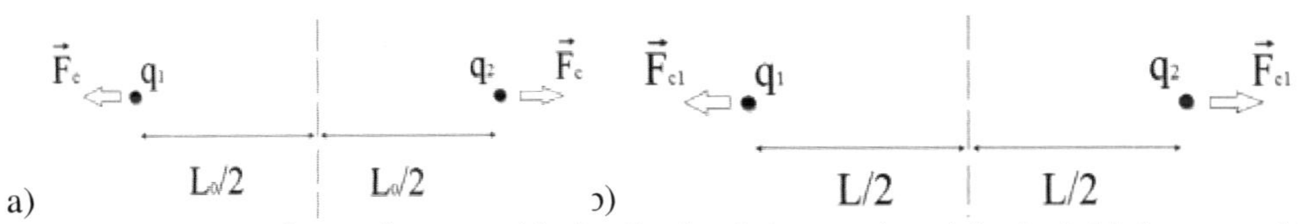

Figure 1. System of two charges with the Coulomb interaction a) in the initial state and b) after the time interval t.

It is well known, that to obtain the average velocity in the time interval t, one should divide the distance traveled by the time taken to travel that distance:

$$\upsilon_{av} = \int_{t_1}^{t_2} \upsilon dt' \bigg/ \int_{t_1}^{t_2} dt' \qquad (1)$$

Then, using the expression for velocity, we obtain υ_{av} in the implicit form:
$\upsilon_{av} = \int_0^t \upsilon dt' \Big/ t = \upsilon_0 + k\dfrac{q^2}{m_0 L_0 t}\left[\dfrac{t}{2\upsilon_{av}} + \dfrac{L_0}{4\upsilon_{av}^2}\ln\left(\dfrac{L_0}{L_0 + 2\upsilon_{av} t}\right)\right]$.

In the case of doubling of distance between the particles to $2L_0$, we have $2L_0 - L_0 = L_0 = \upsilon_{av} t$, and for each one, we obtain the average velocity, expansion time and achieved velocity as $\upsilon_{av_{2x}} = \sqrt{k\dfrac{q^2}{2m_0 L_0}(1-\ln 2)}$, $t_{2x} = \sqrt{\dfrac{m_0 L_0^3}{2kq^2(1-\ln 2)}}$ and $\upsilon_{2x} = \sqrt{k\dfrac{q^2}{8m_0 L_0 (1-\ln 2)}} \approx 0{,}638\sqrt{k\dfrac{q^2}{m_0 L_0}}$, respectively.

One can also obtain the velocity of particles in expansion from the energy conservation equation $k\frac{q^2}{L_0}=k\frac{q^2}{L}+2\frac{mv^2}{2}$. Thus, for the double expansion, we have $v_{2x}=\sqrt{k\frac{q^2}{2m_0L_0}}\approx 0.707\sqrt{k\frac{q^2}{m_0L_0}}$.

The results for velocity values in both approaches are approximately equal, which indicates the correctness of the presented method. However, by considering the interaction at each time moment, the first result corresponds to the more accurate value.

3. Cylindrical volume of the charged particles expanding along the axis

Consider a system of particles with the same charges $q_i=q_{i+1}=q$ and masses m_0 that are densely and uniformly arranged within a cylindrical volume. We denote the particle concentration as n_q. The average distance between the particles is $n_q^{-1/3}$. We find the velocity of the test particle at the surface end of the cylinder extending in a longitudinal direction of the axis (Fig. 2a). We assume that the radial expansion is absent, $R_0=const$. Also, due to the conservation of the number of particles, the total charge in the volume is $q_{total}=qn_qV=qn_q\pi R_0^2L_0=const.$, where V is the volume, R_0 is the initial radius of the cylinder, and L_0 is the initial length of the cylinder. While expanding along the axis to L, for any moment of time according to the Gauss theorem, assuming that $R>>L>L_0$, we obtain the electric field of the entire volume that acts on a test particle on the surface of the cylinder (edge effects are ignored) $E=\frac{qn_qL_0}{2\varepsilon_0}$.

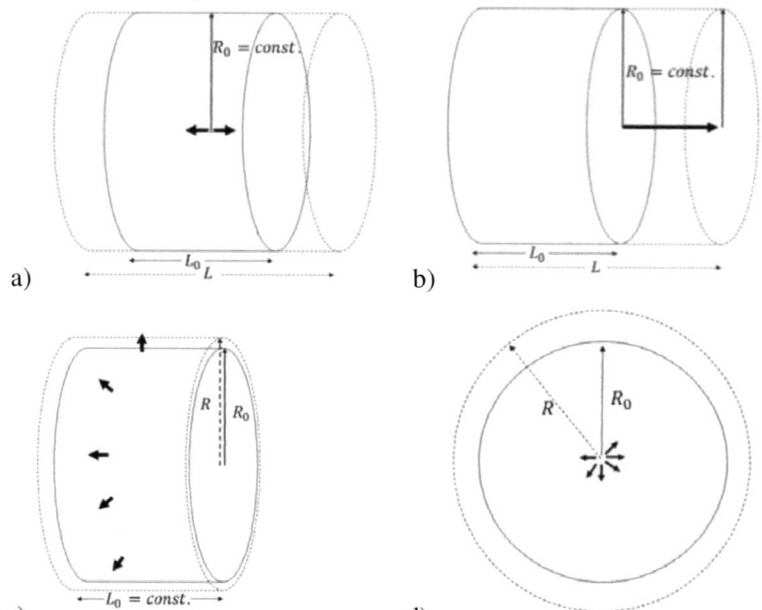

Figure 2. Expansion of the cylindrical volume with similar charged particles along the axis a) in both directions, b) in one direction, c) in radial direction, and d) the expansion in the radial direction of spherical volume with similar charged particles.

The test charge will be acted upon by the force $F = qE = m_0 \frac{d\upsilon}{dt}$. From it, taking $\upsilon_0 = 0$, the expression for the velocity is $\upsilon = \frac{q^2 n_q L_0}{2m_0\varepsilon_0} t$. Let us perform the "linearization" of elongation, assuming that on the interval of cylinder's length change from L_0 to L in time t, the average velocity of the particle is equal to υ_{av}, which strictly corresponds to exactly this interval. Due to the symmetry of both ends of the cylinder, $L - L_0 = 2\upsilon_{av} t$. Using the expression for velocity, (1) and "linearization", we obtain $\upsilon_{av} = \frac{1}{4}\frac{q^2 n_q L_0 t}{m_0\varepsilon_0}$.

For the double expansion to $L = 2L_0$, the average velocity is $\upsilon_{av_{2x}} = \frac{L_0}{2\sqrt{2}}\sqrt{\frac{q^2 n_q}{m_0\varepsilon_0}}$, the time is $t_{2x} = 2\sqrt{\frac{m_0\varepsilon_0}{q^2 n_q}}$, and the velocity is $\upsilon_{2x} = L_0\sqrt{\frac{1}{2}\frac{q^2 n_q}{m_0\varepsilon_0}}$.

Let us consider the same cylindrical volume of the charged particles, but expanding only in one direction along the axis (the second end is fixed). (Fig. 2b). The cylinder expansion times in this and previous problems should be the same because the time is invariant to the reference system (in our non-relativistic case). This can be seen by the example of doubling the length ($L = 2L_0$). We obtain $t_{2x} = 2\sqrt{\frac{m_0\varepsilon_0}{q^2 n_q}}$, which confirms our findings. The velocity is equal to $\upsilon_{2x} = 2L_0\sqrt{\frac{q^2 n_q}{m_0\varepsilon_0}}$. The derivation and assumptions of expressions are similar to the previous problem.

Now, consider the cylindrical volume of the charged particles that are expanding in the radial direction (Fig. 2c). In this case, it is necessary to take $L_0 >> R > R_0$ while expanding in radial direction to R. Neglecting edge effects, for any moment, according to the Gauss theorem for the test particle on the surface of the cylinder, we get $E = \frac{qn_q R_0^2}{2\varepsilon_0 R}$.

Taking $\upsilon_0 = 0$ and "linearizing" as $R - R_0 = \upsilon_{av} t$, the equation for velocity becomes $\upsilon = \frac{q^2 n_q R_0^2}{2m_0\varepsilon_0\upsilon_{av}} \ln\left(\frac{\upsilon_{av} t}{R_0} + 1\right)$. Using this expression and (1), we implicitly obtain the expression for υ_{av}:

$$\upsilon_{av} = \int_0^t \upsilon dt' \Big/ t = \frac{q^2 n_q}{2m_0\varepsilon_0}\frac{R_0^2}{\upsilon_{av} t}\left[\left(\frac{R_0}{\upsilon_{av}} + t\right)\ln\left(\frac{\upsilon_{av} t}{R_0} + 1\right) - t\right].$$

For the double size increase to $R = 2R_0$, we get $\upsilon_{av_{2x}} = R_0\sqrt{\left(\ln 2 - \frac{1}{2}\right)\frac{q^2 n_q}{m_0\varepsilon_0}}$, and then, the time is $t_{2x} = \sqrt{\left(\frac{2}{2\ln 2 - 1}\right)\frac{m_0\varepsilon_0}{q^2 n_q}}$, and the velocity is $\upsilon_{2x} = R_0\sqrt{\left(\frac{1}{4\ln 2 - 2}\right)\frac{q^2 n_q}{m_0\varepsilon_0}}$.

4. Spherical volume of charged particles that are expanding in radial direction

Now, consider a system of identical charged particles that are densely located within a spherical volume. The expansion occurs uniformly in radial direction (Fig. 2d). We will find the velocity of the test particle located on the surface of the sphere in the absence of particle loss

$q_{total} = qn_qV = qn_q\frac{4}{3}\pi R_0^3L_0 = const$. According to the Gauss theorem, $E = \frac{qn_qR_0^3}{3\varepsilon_0R^2}$. Taking $\upsilon_0 = 0$ and "linearizing" as $R - R_0 = \upsilon_{av}t$ for the obtained velocity of the test charge q, we obtain the expression $\upsilon = \frac{q^2n_qR_0^3}{3m_0\varepsilon_0\upsilon_{av}}\left[\frac{1}{R_0} - \frac{1}{R_0 + \upsilon_{av}t}\right]$. Using the expression for velocity and also (1), we implicitly obtain the expression for υ_{av}:

$$\upsilon_{av} = \int_0^t \upsilon dt' \Big/ t = \frac{q^2n_qR_0^2}{3m_0\varepsilon_0\upsilon_{av}} - \frac{q^2n_qR_0^3}{3m_0\varepsilon_0\upsilon_{av}^2t}\ln\left(\frac{\upsilon_{av}t}{R_0} + 1\right).$$

In the case of doubling the size to $R = 2R_0$, we have $\upsilon_{av_{2x}} = R_0\sqrt{\left(\frac{1 - \ln 2}{3}\right)\frac{q^2n_q}{m_0\varepsilon_0}}$. Then, the time is $t_{2x} = \sqrt{\left(\frac{3}{1 - \ln 2}\right)\frac{m_0\varepsilon_0}{q^2n_q}}$, and the velocity is $\upsilon_{2x} = R_0\sqrt{\left(\frac{1}{12(1 - \ln 2)}\right)\frac{q^2n_q}{m_0\varepsilon_0}}$.

5. Generalization of the results

In calculations, if one assumes that the uniform distribution of particles in the corresponding volumes is preserved at any instance of time, the particle velocities inside the volume will be proportional to the distance from the propagation epicenter. In addition, it is necessary to correctly take into account the constraints when using the Gauss theorem in calculations. The calculations for spherical volume assume no restrictions.

The calculation of kinetic energy acquired for the considered test particles in the case of doubling the dimensions is not difficult, in accordance with the expression for kinetic energy $E_{kin} = \frac{1}{2}m\upsilon^2$. However, for the numerical results, it is necessary to set the initial configuration sizes (L_0, R_0). For example, in [2], the authors find the cluster size, which allows to achieve fusion energies when doubling the dimension.

In the case of a system of particles with charges of different sign, but with an uncompensated total charge, the configuration is also subject to an "explosion". In this case, the values of the total uncompensated charge should be used instead of q_{total} in calculations and results obtained in this paper.

6. Conclusions

In this paper, for the first time, the method of "linearization" of the dependence of the dimension parameters on time was applied for calculating the Coulomb "explosion" characteristics. Based on the particle interactions (part 2), the accuracy of the method has been shown. Considering it, the calculations were made for various configurations (parts 3, 4). Analytical formulas for calculating the parameters are obtained. Some results are presented in an implicit form. An explicit form can be obtained by specifying the finite dimension values of the system. Thus, here, taking into account all of the calculations, the results are shown for the double increase in dimensions. These formulas may be used as a basis for the calculation of "explosion" characteristic parameters.

References

[1] Voronchev V T and Kukulin V I 2010 *Yadernaya Fizika* **73** №1 41-61
[2] Zweiback J et al 2000 *Phys. Rev. Lett.* **84** №12 2634-2637
[3] Last I and Jortner J 2006. *Phys. Rev. Lett.* **97** 173401
[4] Savelyev I V **1982** *Kurs Obshchei Fiziki* 1

[5] Savelyev I V **1988** *Kurs Obshchei Fiziki* 2